Phase Space Crystals

Condensed matter in dynamical systems

Phase Space Crystals

Condensed matter in dynamical systems

Lingzhen Guo
School of Science, Nanjing University of Science and Technology, Nanjing, China
and
Max Planck Institute for the Science of Light, Erlangen, Germany

IOP Publishing, Bristol, UK

ISBN 978-0-7503-3563-8 (ebook)
ISBN 978-0-7503-3561-4 (print)
ISBN 978-0-7503-3564-5 (myPrint)
ISBN 978-0-7503-3562-1 (mobi)

DOI 10.1088/978-0-7503-3563-8

Version: 20211201

IOP ebooks

British Library Cataloguing-in-Publication Data: A catalogue record for this book is available from the British Library.

Published by IOP Publishing, wholly owned by The Institute of Physics, London

IOP Publishing, Temple Circus, Temple Way, Bristol, BS1 6HG, UK

US Office: IOP Publishing, Inc., 190 North Independence Mall West, Suite 601, Philadelphia, PA 19106, USA

To my wife Dengxia and our daughter Sarah

Contents

Preface

In summer of the year 2012, I was puzzled by a weird property of the Duffing oscillator and parametric oscillator. The stationary solution of such far-from-equilibrium quantum systems at zero temperature satisfies the detailed balance condition, which is usually considered as a property for equilibrium systems with microscopic reversibility. My colleague, Dr Micheal Marthaler at Karlsruhe Institute of Technology (Germany), suggested checking other highly nonlinearly driven oscillators. I observed a more intriguing phenomenon that the single Floquet band tends to form a sub-band structure. Later on, we found that a hidden discrete rotational symmetry exists in the phase space. We then developed a quasinumber theory analogue to band theory in solid state physics, and called this phenomenon phase space crystals in our paper (2013 *Phys. Rev. Lett.* **111** 205303). We proposed to realise phase space crystals with cold atoms.

However, we didn't know if the interactions between atoms can destroy the band structure. At the same time, I also tried to generalise the discrete symmetry into the whole phase space. In June 2016, during my visit to Heidelberg university, I learned that a kicked harmonic oscillator can generate various lattice structures in phase space. The single-particle physics in such phase space lattice potential has been studied for a long time, e.g., the chaotic diffusion and topological physics. However, it was still unclear what the role of interactions is in such dynamical systems. We later identified that, within the rotating wave approximation, the interactions do not break the discrete or lattice symmetry in phase space. We refined the definition of phase space crystal as the nonequilibrium many-body state existing periodically in the phase space of dynamical systems.

In March 2017, I learned from Professor Krzysztof Sacha at the Jagiellonian University (Poland) that my phase space crystal is closely related to time crystal. Although the time crystal was first proposed by Frank Wilczek in the year 2012, it was then proved impossible for a generic static physical system. Later, Wilczek's idea was extended to periodically driven (Floquet) systems. The concept of time crystals was recused and dubbed as Floquet or discrete time crystals. Floquet time crystals focus on the spontaneous breaking of discrete time-translational symmetry and the stabilising mechanism of subharmonic modes while phase space crystals pay more attention on the interplay of basic response modes. Independently and surprisingly, Krzysztof was working on the same topic of effective interactions in the rotating frame. He called the many-body physics of interacting time crystals the condensed matter physics in time crystals.

As there is a crystal in phase space, it is natural to study its vibrational modes analogue to the phonons in a solid. In January 2018, I joined Professor Florian Marquardt's division at the Max Planck Institute for the Science of Light (Germany). Florian is an expert on the topological sound and light at the nanoscale. When I showed him my preliminary results about the phase space crystal vibrational modes, he immediately pointed out that these modes could have some topological properties. My colleague Dr Vittorio Peano joined my project later. Actually,

Vittorio was one of my collaborators on the detailed balance problem mentioned above. Vittorio found that the topological phonon excitations of phase space crystal have a chiral edge state with preserved time-reversal symmetry, which is surprising because the breaking of time-reversal symmetry is a strict precondition for topological chiral transport in real space.

What is the future of phase space crystals? The general roadmap is to develop the topics analogue to that in condensed matter physics. The physics of phase space crystals also has an intimate relation with the von Neumann lattice, Wigner crystal in magnetic field and quantum Hall physics. For practical applications, phase space lattice potentials can also be used to design bosonic codes for fault-tolerant quantum computing. These interesting topics will be introduced in my book. I hope this book will stimulate more scientists to work on phase space crystals, a young research field with many intriguing phenomena to be explored.

Lingzhen Guo
Bad Endorf, Germany
November 2021

Acknowledgement

I would like to thank my PhD thesis supervisors Prof. Xinqi Li and Prof. Gerd Schön. I am extremely grateful to Gerd's support and encourage during my research at the Karlsruhe Institute of Technology (KIT). I deeply thank Prof. Mark Dykman, Micheal Marthaler and Vittorio Peano for their advice and help on my PhD project. I also thank other colleagues in KIT: Panagiotis Kotetes, Juha Leppäkangas, Stephan André, Juan Atalaya, Jared Cole, Dmitry Golubev, Peiqing Jin, Jinshuan Jin, Roland Grein, Andreas Heimes, Thomas Hellmuth, Regine Frank, Pengfei Liang, Modan Liu, Andreas Poenicke and Sebastian Zanker for helpful discussions. I have happy memories of the years in Karlsruhe.

I would deeply thank Prof. Florian Marquardt for his support on my phase space crystal research at the Max Planck Institute of Light (MPL). It is very inspirational to discuss with Florian as he always provides illuminating insight when I got stuck somewhere. I would also like to thank my colleagues in MPL: Vittorio Peano, Hugo Ribeiro, Tirth Shah, Thomas Fösel, Riccardo Porotti, Leopoldo Sarra, Thales Figueiredo Roque, Carlos Navarrete-Benlloch, Stefan Nimmrichter and Petru Tighineanu for helpful discussions. I also thank our secretary Ms Gesine Murphy in the division.

I also thank Prof. Krzysztof Sacha and Dr Krzysztof Giergiel at the Jagiellonian University (Poland) for helpful discussions on time crystals. Prof. Krzysztof Sacha also provided pertinent comments on the related parts in the book. I would like give special thanks to Prof. Zhigang Zheng at the Huaqiao University (China), who is my Master thesis supervisor and also provided me a lot of help in writing the book.

I acknowledge financial support from the China Scholarship Council, the Visiting Researcher Scholarship from Karlsruhe House of Young Scientist and the Nachwuchsförderprogramm Carl Zeiss Stiftung (0563-2.8/508/2).

Author biography

Lingzhen Guo

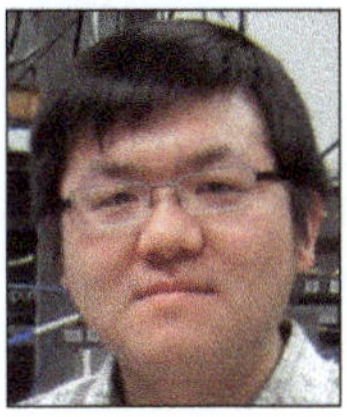

Dr Lingzhen Guo recently took a professor position at the Nanjing University of Science and Technology (NJUST), China. Guo carried out his Bachelor, Master and PhD studies at the Beijing Normal University, China. During the PhD project, Guo spent two years doing joint research at the Karlsruhe Institute of Technology (KIT), Germany. He spent one year of postdoctoral research at KIT and another one year of postdoctoral research at the Chalmers University of Technology, Sweden. After that, he went back to KIT, where he led a two-year independent junior research group funded by Carl-Zeiss-Stiftung. From January 2018 to December 2021, he was a postdoctoral research fellow at the Max Planck Institute for the Science of Light (MPL), Germany. His research interests and activity cover topics in quantum optics, statistical physics and condensed matter physics.

Chapter 1

Introduction

A phase space crystal is a nonequilibrium many-body state appearing in dynamical systems. While solid crystals are defined by a periodic arrangement of atoms in real space, it was recently understood that atoms subject to a time-periodic drive can form crystals in phase space [1]. This is because a periodic drive can produce an effective Hamiltonian with a lattice structure in phase space. The existence of such phase space lattice is guaranteed by the Poincaré–Birkhoff theorem (section 2.2.2). Let us first consider a one-dimensional system *without* periodic driving, e.g., a harmonic oscillator or a nonlinear oscillator. The trajectory of system is a closed isoenergetic contour in phase space with some trajectory frequency. In the presence of a periodic perturbative driving field, the Poincaré–Birkhoff theorem states that, if the ratio of the driving frequency and the trajectory frequency is a rational number, the original regular periodic trajectory will split into equal numbers of stable and unstable trajecttories under any arbitrary small perturbation (section 2.2.2). In the phase space of rotating frame with the driving frequency, the stable points (representing stable trajectories in the rest frame) together with their neighbouring regions form the so called *Poincaré–Birkhoff islands* in the chaotic sea (representing unstable trajectories in the rest frame). These stable regions of Poincaré–Birkhoff islands may form a chain or some regular two-dimensional lattice structures in phase space. Furthermore, the robustness of these Poincaré–Birkhoff islands is guaranteed by the Kolmogorov–Arnold–Moser (KAM) theorem (section 2.2.3).

Historically, the concept of the phase space lattice was first introduced by John von Neumann in his celebrated book *Mathematical Foundations of Quantum Mechanics* [2]. Von Neumann constructed such a phase space lattice by two elementary displacement operators along position and momentum directions respectively. The two commuting operators result in a subset of coherent states that exist periodically in the phase space with a unit cell of area h (the Planck constant), which is nowadays called the *von Neumann lattice*. If the phase space is replaced by a frequency-time plane, the von Neumann lattice is called a *Gabor lattice*

doi:10.1088/978-0-7503-3563-8ch1

[3] and widely used for signal processing [4]. It took quite a long time to prove the existence of a complete and orthogonal basis on the von Neumann lattice [5–10]. The phase space lattice in dynamical systems was investigated by Zaslavsky in his seminal study of kicked harmonic oscillator (KHO) [11]. A KHO is a harmonic oscillator subjected to a periodically kicked cosine potential. Depending on how many kicks during one harmonic period, the effective Hamiltonian in the phase space of rotating frame can possess a square lattice, triangle lattice and even quasi-crystal structures (sections 2.2.5 and 3.1.2).

The phase space lattice differs fundamentally from the real space lattice in the sense that phase space has noncommutative geometry in quantum mechanics. As shown in the left panel of figure 1.1, a coherent state moving along a closed path acquires an additional phase factor $e^{iS/\hbar}$ with S the area of enclosed curve (see section 4.1.2 for more details). This geometric phase is similar to the Aharonov–Bohm phase of a charged particle moving in a magnetic field. For this reason, the physics of phase space lattice has been known to appear naturally in the motion of a charged particle in a two-dimensional plane with a magnetic field [12]. In an out-of-plane magnetic field, the trajectories of charged particles can be decomposed into the cyclotron motion and the drift of guiding centre (magnetron motion). In the limit of the infinite charge-to-mass ratio, the dynamics of guiding centre is reduced to the form of the Lorenz force (see section 4.1.2 for more discussion). In fact, the canonical equation of motion (EOM) can also be cast into the form of Lorenz force in phase space, as shown in the right panel of figure 1.1. Due to the deep connection between phase space and magnetic field, the band structure of a phase space lattice Hamiltonian has an intrinsic topological property that is described by a Chern insulator. The eigenstates with lattice symmetry in phase space are potential candidates for fault-tolerant quantum computing, i.e., the bosonic Gottesman–Kitaev–Preskill (GKP) code (section 3.3.3).

Although the single-particle physics in phase space lattices has been intensively investigated in the past, the many-body dynamics in phase space, which is the central topic of this book, is still lacking study. For the many-body effects, the interaction of particles play a crucial role. It was found that a short-range interaction in real space (e.g., contact interaction of cold atoms) becomes a long-range interaction (e.g., Coulomb-like interaction) in phase space [1, 13–17]. This is because particles that are localised about distant phase space points will still collide in the course of their

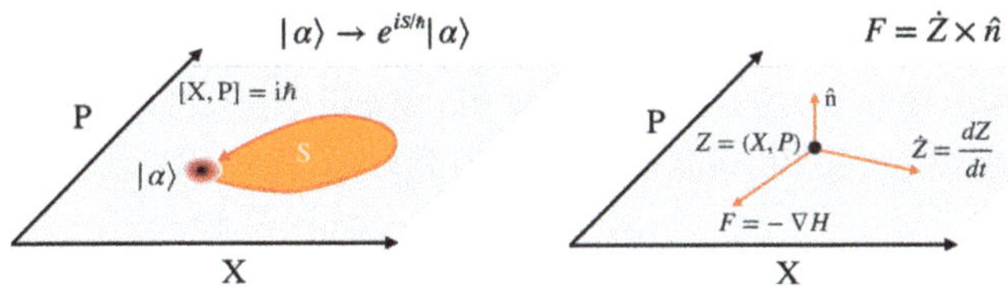

Figure 1.1. Connection between phase space and magnetic field: (left) in quantum mechanics, a coherent state moving along a closed path in phase space acquires a geometric phase proportional to the enclosed area S due to the noncommutative geometry of phase space; (right) in classical mechanics, the canonical equation of motion is cast into the form of Lorenz force by defining a phase space force $F \equiv -\nabla H = -(\partial_X H, \partial_P H)$.

lab-frame trajectories and the effective interaction is basically the averaged interaction over one driving period (section 4.1.1). If particles have spins, the collision of particles can induce an effective spin–spin interaction (section 2.2.2). With the ingredients of phase space lattice potential, effective magnetic field in phase space and the long-range phase space interactions, many strongly-correlated models for electrons in solids, e.g., the Hubbard model and Heisenberg model, can be simulated in phase space lattices (section 4.2.4). It is also possible to introduce disorder to the phase space lattice potential by the imperfect driving field. Thus, the Anderson model and many-body localisation can be simulated in a phase space lattice (section 4.2.4).

The extreme points of a phase space lattice Hamiltonian are the stable points, which are called the *attractors* in dynamical systems. In the presence of dissipation, particles intend to relax to the attractors reaching an equilibrium state between driving (energy pumping) and dissipation (energy loss). If each attractor is occupied by one particle, the periodically arranged particles in phase space form a crystal in phase space, which we call the *phase space crystal.* We point out the difference between a phase space lattice and phase space crystal: *a phase space lattice by our definition is a single-particle Hamiltonian while a phase space crystal is a many-body state.* Similar to what happens in real space, the phase space interactions give rise to collective phonon excitations of phase space crystals. The phonons in phase space are quite different from the phonons in a two-dimensional real space (e.g., graphene). In a solid, the motion of each particle's displacement is governed by a second-order time differential equation. However, for phase space crystals, the motion of each particle is described by a first-order time differential equation, namely, the symplectic dynamics of canonical EOM. Each particle is actually similar to a gyroscope with fixed rotating directions depending on the curvature of their stable point. The combination of symplectic dynamic and interactions in phase space gives a *complex* pairing interaction between vibrations of any two particles (see section 5.1.2 for more details). Their complex phases have a simple geometrical interpretation and cannot be eliminated by a gauge transformation. They can lead to a phonon Chern insulator, accompanied by topologically robust chiral transport along the edge of the phase space crystal. Remarkably, these arise despite the presence of time-reversal symmetry, in contrast to all topological transport scenarios in real space. In this way, phase space crystal vibrations offer a new platform for phononic topological transport, with chiral properties arising naturally (see chapter 5 for detailed discussion).

The concept of phase space crystals is closely related to the concept of Floquet or discrete time crystal (see section 6.3 for more discussion). They both study the subharmonic modes emergent in periodically driven systems. The subharmonic mode responses at a fraction of driving frequency and thus breaks the discrete time translation symmetry (DTTS) of the original time-periodic Hamiltonian. Most studies of Floquet time crystals focus on the spontaneous breaking of DTTS and the protection mechanism of subharmonic modes in quantum many-body systems. All the subharmonic modes have the same frequency but differ from their oscillating amplitudes and phases (or quadratures). The study of phase space crystals focuses more on the periodic pattern and the interplay of potential subharmonic modes of system after DTTS is broken. Actually, the basic modes constructing a phase space crystal can be single-particle modes and need not break DTTS. Phase space crystals focus on the many-body physics of these basic modes with consideration of their interactions. There is a trend to study the interplay of two or more time crystals, e.g., the AC Josephson effect between two superfluid time crystals [18]. The study of many-body physics of many interacting time crystals is coined as *condensed matter physics in time crystals* [16, 19].

There are many works on condensed matter physics with periodic driving field, which are generally called *Floquet engineering* [20]. However, there is a crucial difference of these works from the study of phase space crystals: the condensed matter phenomena by usual Floquet engineering is in the *configuration space* while phase space crystals focus on the condensed matter physics in the *state space* of a dynamical system. The state of a dynamical system is given by a tuple of variables (a state vector) that can be presented by a point in a proper state space, e.g., phase space spanned by positions and momentums. The canonical transformations blur the distinguishment of generalised positions and generalised momenta (section 2.1.3). We can view the phase space as a unified continuum just like spacetime continuum. There is an idea of introducing condensed matter physics in synthetic dimensions by exploiting and harnessing some of the internal degrees of freedom of atoms or photons to simulate motion along extra spatial directions [21–23]. Again, all the synthetic dimensions only enlarge the dimensions of configuration space, and the considered condensed matter physics is still in the configuration space.

One may ask what is the motivation for discussing condensed matter physics in phase space. First of all, it is from pure curiosity since this idea was not previously considered, and the spirit of basic physics studies is to explore new phenomena as long as it is not forbidden by basic laws of physics. Second, the studying of phase space crystals is the natural evolution from the early studies of von Neumann lattice and Poincaré–Birkhoff dynamics to the later studies of stochastic web (section 2.2.5) and recent studies of time crystals (chapter 6). The dynamics in phase space are closely related to the research of the (fractional) quantum Hall effect [24], topological physics [1, 25, 26] and noncommutative geometry [27, 28]. Finally, the study of phase space crystals may have some potential applications, e.g., for the continuous-variable quantum computing with GKP codes (section 3.3.3).

This monograph is organised as follows. In chapter 2, we introduce the basic knowledge of classical dynamical systems and quantum Floquet systems. The classical part (sections 2.1 and 2.2) includes two important results in dynamical systems, i.e., the Poincaré–Birkhoff theorem and the KAM theorem. The two theorems guarantee the existence and robustness of phase space lattices in dynamical systems forming the cornerstones of our study. One can skip this part if the reader, who has the knowledge of canonical mechanics, is not interested in the details of the two theorems. The quantum part (section 2.3) introduces the Floquet theory and Magnus expansion, which is the theory framework of our study. In chapter 3, we discuss various phase space lattice structures (section 3.1), the quantum band theory based on the discrete translational symmetry in phase space (section 3.2) and the topological physics in the noncommutative phase space (section 3.3). In chapter 4, we investigate the classical and quantum many-body dynamics in phase space. Many strongly-correlated modes in solid physics for electrons can be simulated with a phase space lattice (section 4.2.4). In chapter 5, we study the phase space crystal vibrations, which are analogue to the lattice waves (phonons) in a solid crystal. The bands of phase space lattice waves have intrinsic topology with nontrivial Chern numbers and chiral edge states in phase space. Finally, in chapter 6, we review the brief research history of time crystals (sections 6.1 and 6.2) and clarify the relation of time crystals to phase space crystals (section 6.3).

References

[1] Liang P, Marthaler M and Guo L 2018 Floquet many-body engineering: topology and many-body physics in phase space lattices *New J. Phys.* **20** 023043

[2] von Neumann J 1955 *Mathematical Foundations of Quantum Mechanics* (Princeton, NJ: Princeton University Press) pp 406–7

[3] Gabor D 1946 Theory of communication *J. Inst. Electr. Eng.* **93** 429–41

[4] Daubechies I 1990 The wavelet transform, time-frequency localization and signal analysis *IEEE Trans. Inf. Theory* **36** 961–1005

[5] Zak J 1967 Finite translations in solid-state physics *Phys. Rev. Lett.* **19** 1385–7

[6] Perelomov A M 1971 On the completeness of a system of coherent states *Theor. Math. Phys.* **6** 156–64

[7] Wilkins J W, Sullivan D J, Rehr J J and Wilson K G 2010 Phase space Wannier functions in electronic structure calculations arXiv: 1007.3522

[8] Daubechies I, Jaffard S and Journés J-L 1991 A simple Wilson orthonormal basis with exponential decay *SIAM J. Math. Anal.* **22** 554–73

[9] Zak J 2002 Wilson-like functions on von Neumann lattices *J. Phys. A: Math. Gen.* **35** L369–75

[10] Zak J 2003 von Neumann's hypothesis concerning coherent states *J. Phys. A: Math. Gen.* **36** L553–60

[11] Zaslavsky G M 2008 *Hamiltonian Chaos and Fractional Dynamics* (Oxford: Oxford University Press)

[12] Zak J 1992 Identities for landau level orbitals *Europhys. Lett.* **17** 443–8

[13] Sacha K 2015 Anderson localization and Mott insulator phase in the time domain *Sci. Rep.* **5** 10787

[14] Sacha K 2015 Modeling spontaneous breaking of time-translation symmetry *Phys. Rev.* A **91** 033617

[15] Guo L, Liu M and Marthaler M 2016 Effective long-distance interaction from short-distance interaction in a periodically driven one-dimensional classical system *Phys. Rev.* A **93** 053616

[16] Sacha K and Zakrzewski J 2017 Time crystals: a review *Rep. Prog. Phys.* **81** 016401

[17] Giergiel K, Miroszewski A and Sacha K 2018 Time crystal platform: From quasicrystal structures in time to systems with exotic interactions *Phys. Rev. Lett.* **120** 140401

[18] Autti S, Heikkinen P J, Mäkinen J T, Volovik G E, Zavjalov V V and Eltsov V B 2021 AC Josephson effect between two superfluid time crystals *Nat. Mater.* **20** 171–4

[19] Guo L and Liang P 2020 Condensed matter physics in time crystals *New J. Phys.* **22** 075003

[20] Rudner M S and Lindner N H 2020 Band structure engineering and non-equilibrium dynamics in Floquet topological insulators *Nat. Rev. Phys.* **2** 229–44

[21] Celi A, Massignan P, Ruseckas J, Goldman N, Spielman I B, Juzeliunas G and Lewenstein M 2014 Synthetic gauge fields in synthetic dimensions *Phys. Rev. Lett.* **112** 043001
[22] Ozawa T and Price H M 2019 Topological quantum matter in synthetic dimensions *Nat. Rev. Phys.* **1** 349–57
[23] Yuan L, Lin Q, Xiao M and Fan S 2018 Synthetic dimension in photonics *Optica* **5** 1396–405
[24] Ishikawa K and Maeda N 1997 Flux state in von Neumann lattices and the fractional hall effect *Prog. Theor. Phys.* **97** 507–25
[25] Leboeuf P, Kurchan J, Feingold M and Arovas D P 1990 Phase-space localization: topological aspects of quantum chaos *Phys. Rev. Lett.* **65** 3076–9
[26] Leboeuf P, Kurchan J, Feingold M and Arovas D P 1992 Topological aspects of quantum chaos *Chaos* **2** 125–30
[27] Connes A 1994 *Noncommutative Geometry* (New York: Academic)
[28] Bellissard J, van Elst A and Schulz-Baldes H 1994 The noncommutative geometry of the quantum hall effect *J. Math. Phys.* **35** 5373–51

Chapter 2

Dynamical systems

In this chapter, we introduce the knowledge of dynamical systems for phase space crystals. We start to summarise briefly the basics of canonical dynamics in section 2.1. Then, in section 2.2, we discuss near-integrable systems and weakly-chaotic dynamics. A generic non-integrable dynamical system exhibits chaotic behaviour together with regular motions coexisting in phase space. This book is not aimed at discussing chaotic dynamics but focuses on the regular regions, which are called Poincaré–Birkhoff islands, in phase space. We introduce two important theorems of dynamical systems: the Poincaré–Birkhoff theorem and the Kolmogorov–Arnold–Moser (KAM) theorem. The two theorems guarantee the existence and robustness of phase space lattices and form the cornerstones of phase space crystals. In section 2.3, we introduce the Floquet theory and Magnus expansion for quantum systems, which provide the basic framework for the discussion of phase space crystals throughout this book.

2.1 Canonical mechanics

2.1.1 Lagrangian mechanics

In Newtonian mechanics, a *point mass* refers to a point particle with mass but no internal structure. It needs 3 real numbers to label the position of a point mass in three-dimensional space. For a system composed of N point masses, the positions of all the point masses are labelled by a set of $3N$ Cartesian coordinates $\{x_i(t), i = 1, 2, \dots, 3N\}$, which is called the *d'Alembert configuration*. From Newton's laws of motion, we can directly build $3N$ equations of motion describing the time evolution of all the Cartesian coordinates. However, there are possible constraints on the point masses of the system which, for example, determine the shape of an extended body. Usually, it is very difficult or even impossible to solve the set of $3N$ equations of motion. Therefore, it is better to minimise the total number of coordinates by considering the constraints. With the help of symmetry analysis and other methods, we can reduce the total $3N$ Cartesian coordinates into n independent *generalised coordinates* $\{q_i(t), i = 1, 2, \dots, n = 3N - Nc\}$, where N_c is the number

doi:10.1088/978-0-7503-3563-8ch2

of constraints. Each n-tuple of generalised coordinates specifies a possible configuration of the system, which is called the *Lagrange configuration.* The set of all the Lagrange configurations is called the *configuration space* of the system. For systems with integrable constraints, the dimension of the configuration space equals the number of *degrees of freedom* of the system [1].

In configuration space, Newtonian mechanics can be reformed into the so-called *Lagrangian mechanics.* We define the n-tuple $\boldsymbol{q} = (q_1, q_2, \dots, q_n)$ as the position vector of the system in the configuration space. Consider the system starts from a point A at initial time instant $t = t_1$ and finally reaches a point B at later time instant $t = t_2$. Between points A and B, we can draw an infinite number of paths parameterised by time $\boldsymbol{q}(t)$. In order to determine the real dynamics of the system, we define the *Hamiltonian action* S as the integral of an undetermined function, i.e., the *Lagrangian* $L(\boldsymbol{q}, \dot{\boldsymbol{q}}, t)$, between A and B

$$S[\boldsymbol{q}(t)] \equiv \int_{t_1}^{t_2} L(\boldsymbol{q}, \dot{\boldsymbol{q}}, t)dt, \tag{2.1}$$

which is a functional of path $\boldsymbol{q}(t)$. Here, the Lagrangian $L(\boldsymbol{q}, \dot{\boldsymbol{q}}, t)$ is a function of generalised coordinates $\boldsymbol{q}$, *generalised velocities* $\dot{\boldsymbol{q}}$ and time t. For a given Lagrangian, the real path of the system in configuration space is determined by *Hamilton's principle* or the *least action principle*

$$\delta S[\boldsymbol{q}(t)] \equiv \delta \int_{t_1}^{t_2} L(\boldsymbol{q}, \dot{\boldsymbol{q}}, t)dt = 0. \tag{2.2}$$

Here, δS means the variation of action S due to the variation of path $\boldsymbol{q}(t)$. We assume an infinitely small virtual displacement of $\boldsymbol{q}(t)$, i.e., $\boldsymbol{q}(t) \to \boldsymbol{q}(t) + \delta\boldsymbol{q}(t)$, with constraints at the two endpoints (A and B) $\delta\boldsymbol{q}(t_1) = \delta\boldsymbol{q}(t_2) = 0$. The variation of action is given by

$$\delta S = \int_{t_1}^{t_2} dt \sum_{i=1}^{n}\left(\frac{\partial L}{\partial q_i}\delta q_i + \frac{\partial L}{\partial \dot{q}_i}\delta \dot{q}_i\right). \tag{2.3}$$

Here, we take the equal-time variations and the variation operator δ commutes with the time-differential operator d/dt, i.e.,

$$\delta \dot{q}_i(t) = \delta\frac{dq_i}{dt} = \frac{d}{dt}\delta q_i. \tag{2.4}$$

Using the above identity, we can write the variation of action in the following form

$$\delta S = \left(\sum_{i=1}^{n}\frac{\partial L}{\partial \dot{q}_i}\delta q_i\right)\Bigg|_{t_1}^{t_2} + \int_{t_1}^{t_2} dt \sum_{i=1}^{n}\delta q_i\left[\frac{\partial L}{\partial q_i} - \frac{d}{dt}\left(\frac{\partial L}{\partial \dot{q}_i}\right)\right]. \tag{2.5}$$

The first term on the right-hand side of the above equation is zero due to the constraints at the two endpoints. The least action principle requires that the left-hand side is zero $\delta S = 0$. Because the variation δq_i is arbitrary, we obtain *Lagrange's equation*

$$\frac{d}{dt}\left(\frac{\partial L}{\partial \dot{q}_i}\right) - \frac{\partial L}{\partial q_i} = 0, \quad i = 1, 2, \dots, n. \tag{2.6}$$

Note that the Lagrangian of a dynamical system is not unique. In fact, two Lagrangians describe the same dynamical system if they only differ by the total time derivative of a function $f(\boldsymbol{q}, t)$, i.e., $L(\boldsymbol{q}, \dot{\boldsymbol{q}}, t) \to L(\boldsymbol{q}, \dot{\boldsymbol{q}}, t) + df(\boldsymbol{q}, t)/dt$, which can be directly verified from Lagrange's equations using $df(\boldsymbol{q}, t)/dt = \partial f(\boldsymbol{q}, t)/\partial t + \sum_{i=1}^{n} \dot{q}_i \partial f(\boldsymbol{q}, t)/\partial q_i$.

For a given dynamical system, the choice of Lagrangian $L(\boldsymbol{q}, \dot{\boldsymbol{q}}, t)$ has to be consistent with Newtonian mechanics. In other words, Lagrange's equations must recover Newton's equations when the Lagrange configuration is transformed back to the d'Alembert configuration. For the systems with conservative forces, the Lagrangian takes the form of kinetic energy minus potential energy, i.e., $L(\boldsymbol{q}, \dot{\boldsymbol{q}}, t) = T(\dot{\boldsymbol{q}}) - U(\boldsymbol{q})$. However, the Lagrangian is not always separated into such form. For example, the Lagrangian of a particle with charge q and mass m in the electromagnetic field is $L(\boldsymbol{x}, \dot{\boldsymbol{x}}, t) = \frac{1}{2}m\dot{\boldsymbol{x}}^2 + q\dot{\boldsymbol{x}} \cdot \boldsymbol{A} - q\phi$. Here, we simply take the position vector $\boldsymbol{x} = (x_1, x_2, x_3)$ in three dimensional space as the configuration space, $\phi(\boldsymbol{x}, t)$ is the electric scalar potential, and $\boldsymbol{A}(\boldsymbol{x}, t)$ is the magnetic vector potential. In the study of elementary particle physics, the Lagrangian is sometimes conjectured from more general symmetry analysis, e.g., the local gauge invariance, and the equations of motion are derived from Lagrange's equations. The correctness of the chosen Lagrangian is finally verified and supported by experiments.

Lorentz equation of motion It can be chcked that Lagrange's equations with the Lagrangian $L(\boldsymbol{x}, \dot{\boldsymbol{x}}, t) = \frac{1}{2}m\dot{\boldsymbol{x}}^2 + q\dot{\boldsymbol{x}} \cdot \boldsymbol{A} - q\phi$ recover the Lorentz equation of motion, i.e.,

$$m\ddot{\boldsymbol{x}} = q\boldsymbol{E} + q\dot{\boldsymbol{x}} \times \boldsymbol{B}, \tag{2.7}$$

where $\boldsymbol{E} = -\nabla\phi - \partial \boldsymbol{A}/\partial t$ is the electric field and $\boldsymbol{B} = \nabla \times \boldsymbol{A}$ is the magnetic field. The two terms in Lagrange's equations are given by

$$\frac{d}{dt}\frac{\partial L}{\partial \dot{x}_i} = m\frac{d\dot{x}_i}{dt} + q\frac{dA_i}{dt} = m\frac{d\dot{x}_i}{dt} + q\sum_{j=1}^{3} \dot{x}_j \frac{\partial A_i}{\partial x_j} + q\frac{\partial A_i}{\partial t}$$

$$\frac{\partial L}{\partial x_i} = -q\frac{\partial \phi}{\partial x_i} + q\sum_{j=1}^{3} \dot{x}_j \frac{\partial A_j}{\partial x_i}.$$

Therefore, we have

$$m\ddot{x}_i = -q\left(\frac{\partial \phi}{\partial x_i} + \frac{\partial A_i}{\partial t}\right) + q\sum_{j=1}^{3} \dot{x}_j \left(\frac{\partial A_j}{\partial x_i} - \frac{\partial A_i}{\partial x_j}\right), \tag{2.8}$$

which is just the x_i-component of the Lorentz equation of motion (2.7).

2.1.2 Hamiltonian mechanics

Lagrange's equations are the real-time second-order differential equations in configuration space. The solution of Lagrange's equations is a trajectory in n dimensional configuration space. There are infinite trajectories passing through the same point. In other words, the future of the dynamical system is not determined by the point in configuration space. Sometimes, this non-uniqueness brings inconvenience to analyse the behaviours of dynamical systems. To circumvent this problem, we can extend the n dimensional configuration space into a $2n$ dimensional *state space*, and reform the Lagrange's equations into real-time first-order differential equations. In this way, the time evolution of the dynamical systems is uniquely determined by the current state (a single point) in state space. Mathematically, this procedure is implemented by the so-called *Legendre transformation*, which is composed of three steps as follows.

1. We define the *generalised momentum* by

$$p_i \equiv \frac{\partial}{\partial \dot{q}_i} L(\boldsymbol{q}, \dot{\boldsymbol{q}}, t), \quad i = 1, 2, \ldots, n. \tag{2.9}$$

   The above set of n equations defines the relation between n generalised velocities $\dot{q}_i$ and n generalised momenta p_i in the form of implicit functions.
2. We calculate the inverse functions of equations (2.9) and express each generalised velocity $\dot{q}_i$ with n defined generalised momenta $\boldsymbol{p} = (p_1, p_2, \ldots, p_n)$ and generalised coordinates $\boldsymbol{q} = (q_1, q_2, \ldots, q_n)$.
3. We define the *conjugate* function of $L(\boldsymbol{q}, \dot{\boldsymbol{q}}, t)$ in the form of

$$H(\boldsymbol{q}, \boldsymbol{p}, t) = \sum_{i=1}^{n} p_i \dot{q}_i - L(\boldsymbol{q}, \dot{\boldsymbol{q}}, t), \tag{2.10}$$

   where $\dot{q}_i(\boldsymbol{q}, \boldsymbol{p}, t)$ must be understood as the inverse functions of equations (2.9).

The generalised momentum p_i is the conjugate variable of q_i. The $2n$ variables (q_i, p_i) representing the state of system are called *canonical variables*. The canonical variables $(\boldsymbol{q}, \boldsymbol{p})$ constitute a state space, which is called the *phase space* of the dynamical system. The function $H(\boldsymbol{q}, \boldsymbol{p}, t)$ is called *Hamilton's function* or simply the *Hamiltonian*.

In order to derive the equation of motion for the system's state in phase space, we start from the total differential of the Lagrangian

$$dL = \sum_{i=1}^{n} \left(\frac{\partial L}{\partial q_i} dq_i + \frac{\partial L}{\partial \dot{q}_i} d\dot{q}_i \right) + \frac{\partial L}{\partial t} dt = \sum_{i=1}^{n} (\dot{p}_i dq_i + p_i d\dot{q}_i) + \frac{\partial L}{\partial t} dt. \tag{2.11}$$

In the second equality, we have used $p_i = \partial L / \partial \dot{q}_i$ and Lagrange's equations (2.6). Using differential identity $p_i d\dot{q}_i = d(p_i \dot{q}_i) - \dot{q}_i dp_i$, we can reform equation (2.11) as following

$$d\left(\sum_{i=1}^{n} p_i \dot{q}_i - L \right) = \sum_{i=1}^{n} (-\dot{p}_i dq_i + \dot{q}_i dp_i) - \frac{\partial L}{\partial t} dt. \tag{2.12}$$

Comparing the above equation to the total differential form of Hamiltonian's function

$$dH = \sum_{i=1}^{n}\left(\frac{\partial H}{\partial q_i}dq_i + \frac{\partial H}{\partial p_i}dp_i\right) + \frac{\partial H}{\partial t}dt, \tag{2.13}$$

we obtain *Hamilton's equations*

$$\dot{p}_i = -\frac{\partial H}{\partial q_i}, \qquad \dot{q}_i = \frac{\partial H}{\partial p_i}, \qquad i = 1, 2, \ldots, n. \tag{2.14}$$

Note that Hamilton's equations are $2n$ first-order differential equations in phase space while Lagrange's equations consist of n second-order equations in configuration space. Due to the simplicity and symmetric form, Hamilton's equations are also called *canonical equations.* Comparing the last terms in equations (2.12) and (2.13), we obtain another relation $\partial H/\partial t = -\partial L/\partial t$, which is also replaced by an alternative form [2]

$$\frac{dH}{dt} = \frac{\partial H}{\partial t} + \sum_{i=1}^{n}\left(\frac{\partial H}{\partial q_i}\dot{q}_i + \frac{\partial H}{\partial p_i}\dot{p}_i\right) = \frac{\partial H}{\partial t}. \tag{2.15}$$

Here, we have used the canonical equations. For a time-independent Hamiltonian, we have $dH/dt = 0$ meaning the energy conservation.

Poisson brackets For two given smooth functions $F(\boldsymbol{q}, \boldsymbol{p}, t)$ and $G(\boldsymbol{q}, \boldsymbol{p}, t)$, we define their *Poisson bracket* by

$$\{F, G\} \equiv \sum_{i=1}^{n}\left(\frac{\partial F}{\partial q_i}\frac{\partial G}{\partial p_i} - \frac{\partial F}{\partial p_i}\frac{\partial G}{\partial q_i}\right). \tag{2.16}$$

The Poisson bracket satisfies the following rules for any three functions F, G and K [2]:

- Antisymmetric relation

$$\{F, G\} = -\{G, F\}. \tag{2.17}$$

- Bilinearity

$$\{aF + bG, K\} = a\{F, K\} + b\{G, K\}, \quad a, b \in \mathbb{R}. \tag{2.18}$$

- Leibnitz's rule

$$\{FG, K\} = F\{G, K\} + \{F, K\}G, \quad \frac{\partial}{\partial t}\{F, G\} = \left\{\frac{\partial F}{\partial t}, G\right\} + \left\{F, \frac{\partial G}{\partial t}\right\}. \tag{2.19}$$

- Jacobi identity

$$\{F, \{G, K\}\} + \{G, \{K, F\}\} + \{K, \{F, G\}\} = 0. \tag{2.20}$$

For the special case $F = q_i$ and $G = p_i$, we have

$$\{q_i, q_j\} = \{p_i, p_j\} = 0, \qquad \{q_i, p_j\} = \delta_{ij}. \tag{2.21}$$

2.1.2.1 *Integral of motion*

For a Hamiltonian system, we can construct a smooth function $F(\boldsymbol{q}, \boldsymbol{p}, t)$ in phase space. As dynamical variables $(\boldsymbol{q}, \boldsymbol{p})$ evolve in time, the values of $F(\boldsymbol{q}, \boldsymbol{p}, t)$ also change with time. We can calculate the total time derivative of F from Hamilton's equations

$$\frac{dF}{dt} = \frac{\partial F}{\partial t} + \sum_{i=1}^{n}\left(\dot{q}_i\frac{\partial F}{\partial q_i} + \dot{p}_i\frac{\partial F}{\partial p_i}\right) = \frac{\partial F}{\partial t} + \sum_{i=1}^{n}\left(\frac{\partial F}{\partial q_i}\frac{\partial H}{\partial p_i} - \frac{\partial F}{\partial p_i}\frac{\partial H}{\partial q_i}\right). \tag{2.22}$$

Using the Poisson bracket defined via equation (2.16), we can rewrite the above time derivative in a simplified form

$$\frac{dF}{dt} = \frac{\partial F}{\partial t} + \{F, H\}. \tag{2.23}$$

We call F an *integral of motion* if it keeps constant when the system evolves in time, i.e., $dF/dt = \partial F/\partial t + \{F, H\} = 0$. If the function F does not explicitly depend on time, the condition that $F(\boldsymbol{q}, \boldsymbol{p})$ is an integral of motion becomes $\{F, H\} = 0$. Given two functions F and G, we can also calculate the time derivative of their Poisson bracket

$$\begin{aligned}\frac{d}{dt}\{F, G\} &= \frac{\partial}{\partial t}\{F, G\} + \{\{F, G\}, H\}\\ &= \left\{\frac{\partial F}{\partial t}, G\right\} + \left\{F, \frac{\partial G}{\partial t}\right\} - \{\{G, H\}, F\} - \{\{H, F\}, G\}\\ &= \left\{\frac{\partial F}{\partial t} + \{F, H\}, G\right\} + \left\{F, \frac{\partial G}{\partial t} + \{G, H\}\right\}.\end{aligned} \tag{2.24}$$

Here, we have used the Leibnitz's rule (2.17) and Jacobi identity (2.20) of the Poisson bracket. According to equation (2.24), the Poisson bracket $\{F, G\}$ is an integral of motion if F and G are both integrals of motion. This conclusion is called *Poisson's theorem*. However, the constructed function $\{F, G\}$ may be meaningless as the number of integrals of motion for a given system is no more than $2n - 1$, where n is the number of degrees of freedom. The Poisson bracket $\{F, G\}$ gives new integral of motion only if it is neither a constant nor a function of F and G [2]. We call functions F and G *in involution* if they satisfy [3]

$$\{F, G\} = 0. \tag{2.25}$$

2.1.2.2 Non-autonomous Hamiltonian systems

A dynamical system is called *non-autonomous* if its Hamiltonian is time-dependent, otherwise it is an *autonomous* Hamiltonian system. For a non-autonomous Hamiltonian system, the energy is not conserved, i.e., $dH/dt = \partial H/\partial t \neq 0$. In fact, an n dimensional non-autonomous Hamiltonian system can be mapped to a $n+1$ dimensional autonomous Hamiltonian system by increasing the number of dimensions. As a result, the conclusions of autonomous Hamiltonian dynamics also apply for the non-autonomous Hamiltonian systems in the extended configuration space.

Consider a non-autonomous system with Hamiltonian $H(\boldsymbol{q}, \boldsymbol{p}, t)$. We introduce an additional coordinate q_{n+1} to replace the time t, which satisfies

$$\frac{dq_{n+1}}{dt} = \frac{dt}{dt} = 1.$$

Accordingly, we introduce a new generalised momentum p_{n+1} conjugate to q_{n+1} associated with the following new Hamiltonian

$$\bar{H}(\boldsymbol{q}, q_{n+1}, \boldsymbol{p}, p_{n+1}) = H(\boldsymbol{q}, \boldsymbol{p}, q_{n+1}) + p_{n+1}, \tag{2.26}$$

which satisfies the following conditions

$$\frac{\partial \bar{H}}{\partial q_i} = \frac{\partial H}{\partial q_i}, \quad \frac{\partial \bar{H}}{\partial p_i} = \frac{\partial H}{\partial p_i}, \quad \frac{\partial \bar{H}}{\partial p_{n+1}} = 1. \tag{2.27}$$

According to the requirement of canonical equations, we have

$$\frac{dp_{n+1}}{dt} = -\frac{\partial \bar{H}}{\partial q_{n+1}} = -\frac{\partial H}{\partial t} = -\frac{dH}{dt}, \tag{2.28}$$

which means $\dot{\bar{H}} = \dot{H} + \dot{p}_{n+1} = 0$. The new Hamiltonian $\bar{H}$ is a conserved quantity and thus a new integral of motion. By choosing the initial condition $p_{n+1}(t_0) = -H(\boldsymbol{q}(t_0), \boldsymbol{p}(t_0), t_0)$, we have $p_{n+1}(t) = -H(\boldsymbol{q}, \boldsymbol{p}, t)$. The conclusions from autonomous systems can be extended to non-autonomous systems by counting the additional pair of conjugate variables $(q_{n+1}, p_{n+1}) = (t, -H)$ together with the new Hamiltonian (2.26). For example, the Poisson bracket equation (2.23) describing the time evolution of a function F can be reformed as

$$\frac{dF}{dt} = \{F, \bar{H}\}. \tag{2.29}$$

The tuple of dynamical variables $(\boldsymbol{q}, \boldsymbol{p}, t)$ consists of a $2n+1$ dimensional state space, which is also called the *extended phase space*.

In figure 2.1(a), we assume a closed contour γ_0 embedding in the extended phase space as the initial condition of canonical equations. All the trajectories emanating

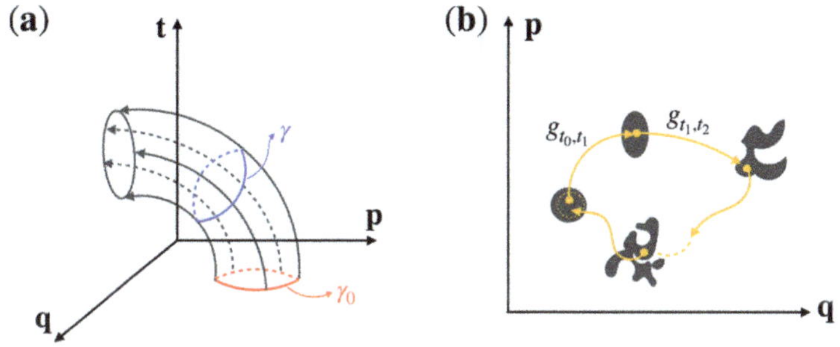

Figure 2.1. (a) Poincaré–Cartan invariant. The trajectories emanating from the initial contour γ_0 form a smooth surface that is called the tube of trajectories. The Poincaré–Cartan integral (2.30) defined on an arbitrary closed contour γ is an invariant. (b) Liouville's theorem. An ensemble of initial states occupying a finite phase space region evolves with distorted shape but preserved phase space volume. g_{t_i,t_j} represents the phase flow between two different time instants.

from the initial contour γ_0 form a smooth curved surface Γ, which is called the *tube of trajectories*. For an arbitrary closed contour γ (surrounding the tube) on the surface Γ, one can prove that the following integral

$$I(\gamma) = \oint_\gamma \left(\sum_{i=1}^{n} p_i dq_i + p_{n+1} dq_{n+1} \right) = \oint_\gamma \left(\sum_{i=1}^{n} p_i dq_i - Hdt \right) \tag{2.30}$$

is an invariant, i.e., $I(\gamma) = I(\gamma_0)$. The integral defined by the above equation is called the *Poincaré–Cartan invariant* [4].

2.1.2.3 Liouville's theorem

In figure 2.1(b), we consider an ensemble of initial states occupying a finite region in phase space. As the Hamiltonian system evolves, the shape of the occupied region may be distorted but the area of the phase space volume is preserved. This important result for Hamiltonian dynamics is called *Liouville's theorem.*

In order to prove Liouville's theorem, we introduce a position vector z in $2n$ dimensional phase space as following

$$z \equiv (\boldsymbol{q}, \boldsymbol{p})^T = (q_1, q_2, \ldots, q_n, p_1, p_2, \ldots, p_n)^T, \tag{2.31}$$

and the gradient operator in phase space

$$\nabla_z \equiv \left(\frac{\partial}{\partial z_1}, \frac{\partial}{\partial z_2}, \ldots, \frac{\partial}{\partial z_{2n}} \right)^T = \left(\frac{\partial}{\partial q_1}, \ldots, \frac{\partial}{\partial q_n}, \frac{\partial}{\partial p_1} \ldots, \frac{\partial}{\partial p_n} \right)^T. \tag{2.32}$$

We write the canonical equations (2.14) in the following compact form

$$\frac{dz}{dt} = \boldsymbol{\Omega} \nabla_z H(z, t) \equiv \boldsymbol{F}(z, t), \tag{2.33}$$

where $\boldsymbol{\Omega}$ is the *symplectic matrix* defined by

$$\boldsymbol{\Omega} = \begin{pmatrix} 0 & \boldsymbol{I}_n \\ -\boldsymbol{I}_n & 0 \end{pmatrix} \tag{2.34}$$

with $\boldsymbol{I}_n$ the $n \times n$ identity matrix. Therefore, the Hamiltonian system has symplectic structure in phase space. The function $\boldsymbol{F}(z, t)$ is a vector field in the $2n$ dimensional phase space and defines a map g from the phase space to itself:

$$g_{t_0,t}: z(t_0) \to z(t_0 + t). \tag{2.35}$$

This map is called *phase flow* of the Hamiltonian system in phase space. It is not difficult to prove the divergence of the vector field $\boldsymbol{F}$ is zero, i.e.,

$$\nabla_z \cdot \boldsymbol{F} = \sum_{i=1}^{n}\left(\frac{\partial \dot{q}_i}{\partial q_i} + \frac{\partial \dot{p}_i}{\partial p_i}\right) = \sum_{i=1}^{n}\left(\frac{\partial^2 H}{\partial q_i \partial p_i} - \frac{\partial^2 H}{\partial p_i \partial q_i}\right) = 0. \tag{2.36}$$

Consider an initial finite region $R(t_0)$ in phase space, for a time interval Δt, the phase flow maps $R(t_0)$ to another region $R(t_0 + \Delta t)$:

$$g_{t_0,\Delta t}(R(t_0)) = R(t_0 + \Delta t). \tag{2.37}$$

We calculate the volume $V(t_0 + \Delta t)$ of the evolved region $R(t_0 + \Delta t)$ as following

$$V(t_0 + \Delta t) = \int_{R(t_0+\Delta t)} \prod_{i=1}^{2n} dz_i = \int_{g_{t_0,\Delta t}(R(t_0))} \prod_{i=1}^{2n} dz_i = \int_{R(t_0)} \mathrm{Jac}(g_{t_0,\Delta t}) \prod_{i=1}^{2n} dz_i, \tag{2.38}$$

where $\mathrm{Jac}(g_{t_0,\Delta t})$ is the Jacobian of the mapping $g_{t_0,\Delta t}$, i.e., the determinant of the derivative of the mapping $g_{t_0,\Delta t}$ with respect to phase space coordinates. For small Δt, we have

$$g_{t_0,\Delta t}(z(t_0)) = z(t_0) + \Delta t \boldsymbol{F}(z(t_0), t_0) + o(\Delta t^2), \tag{2.39}$$

and its derivative with respect to phase space coordinates

$$\nabla g_{t_0,\Delta t} = \boldsymbol{I}_{2n} + \Delta t \nabla_z \boldsymbol{F} + o(\Delta t^2), \tag{2.40}$$

where $\nabla_z \boldsymbol{F}$ is defined via $(\nabla_z \boldsymbol{F})_{ij} = \partial F_i / \partial z_j$. Using the fact that $\det(\boldsymbol{I} + \epsilon \boldsymbol{A}) = 1 + \epsilon \mathrm{tr} \boldsymbol{A} + o(\epsilon^2)$ for the square matrix $\boldsymbol{A}$, we obtain the Jacobian of $g_{\Delta t}$ for a small time interval

$$\begin{aligned} \mathrm{Jac}(g_{t_0,\Delta t}) &= 1 + \Delta t \mathrm{tr}(\nabla_z \boldsymbol{F}) + o(\Delta t^2) = 1 + \Delta t \nabla_z \cdot \boldsymbol{F} + o(\Delta t^2) \\ &= 1 + o(\Delta t^2). \end{aligned} \tag{2.41}$$

Here, we have used equation (2.36). From equation (2.38), we have $V(t_0 + \Delta t) = V(t_0) + o(\Delta t^2)$ meaning the rate of change of the volume is zero $dV/dt = 0$. In other words, the phase flow is incompressible in phase space.

Note that, as the Jacobian of $g_{t0,\Delta t}$ is '+1' for $\Delta t \to 0$, Liouville's theorem also claims the *sign* of volume is preserved. Liouville's theorem holds both in autonomous and non-autonomous Hamiltonian systems.

Poincaré recurrence theorem There is a direct and remarkable corollary from Liouville's theorem which states:

For the Hamiltonian phase flow, the trajectory starting from a point in an arbitrary small neighbourhood $z(0) \in U$ *eventually returns back to the initial neighbourhood* $g_t(z(0)) \in U$.

This corollary is called the *Poincaré recurrence theorem* [1]. In figure 2.1(b), we show a representative trajectory (yellow curves with arrows) emanating from and returning back to a small neighbourhood (enclosed by the yellow dashed circle). Strictly speaking, the Poincaré recurrence holds for a bounded domain in phase space, which is satisfied for a finite-size physical system whose energy cannot increase infinitely. A brief proof of Poincaré recurrence theorem can be made as follows. For a small time interval Δt, we consider a sequence of discrete mapping regions $(g_{\Delta t})^n(U)$ with $n = 0, 1, 2, \ldots$. The successive mapping regions $(g_{\Delta t})^n(U)$ either intersect or not. If they do not intersect, the total volume becomes infinite, which is impossible as the accessible domain in phase space is bounded. If two mapping regions $(g_{\Delta t})^i(U)$ and $(g_{\Delta t})^j(U)$ intersect ($i < j$), then their previous regions $(g_{\Delta t})^{i-1}(U)$ and $(g_{\Delta t})^{j-1}(U)$ must also intersect since a point in the intersection belongs to both sets. Thus, the initial region $(g_{\Delta t})^0(U) = U$ intersects mapping region $(g_{\Delta t})^{j-i}(U)$.

2.1.3 Canonical transformation

According to Hamilton's principle given by equation (2.2), the form of Lagrange's equations (2.6) is independent of the coordinates transformation $(\boldsymbol{q}, t) \to (\boldsymbol{Q}, t)$. This point transformation in configuration space also keeps the form of Hamilton's equations (2.14) unchanged. Since Hamilton's equations are defined in phase space, one may ask if it is possible to perform such a point transformation of phase space coordinates $(\boldsymbol{q}, \boldsymbol{p}, t) \to (\boldsymbol{Q}, \boldsymbol{P}, t)$ that preserves the form of Hamilton's equations. To clarify this problem, let us consider a point transformation in phase space

$$Q_i = Q_i(\boldsymbol{q}, \boldsymbol{p}, t), \quad P_i = (\boldsymbol{q}, \boldsymbol{p}, t). \tag{2.42}$$

Our task is to search for the so-called *canonical transformations* that satisfy

$$\dot{P}_i = -\frac{\partial H'}{\partial Q_i}, \qquad \dot{Q}_i = \frac{\partial H'}{\partial P_i}, \qquad i = 1, 2, \ldots, n. \tag{2.43}$$

Here, $H'(\boldsymbol{Q}, \boldsymbol{P}, t)$ is a new Hamiltonian from the canonical transformation.

There is a general approach to obtain such canonical transformations based on the method of generating functions. In both old and new phase spaces, the system has to satisfy Hamilton's principle, that is to say,

$$\delta \int \left(\sum_{i=1}^{n} p_i \dot{q}_i - H \right) dt = 0, \quad \delta \int \left(\sum_{i=1}^{n} P_i \dot{Q}_i - H' \right) dt = 0. \tag{2.44}$$

Note that we have rewritten Hamilton's principle in a modified form. In the modified Hamilton's principle (2.44), phase space coordinates (q_i, p_i) and (Q_i, P_i) are both independent variables. Both variations with two different coordinates need to hold for the same dynamical system. The two integrands differ at most by a total differential of a function G,

$$\sum_{i=1}^{n} p_i dq_i - H dt = \sum_{i=1}^{n} P_i dQ_i - H' dt + dG. \tag{2.45}$$

The function G is called the *generating function* which is, in the most general form, a function of coordinates $\boldsymbol{q}$, $\boldsymbol{p}$, $\boldsymbol{Q}$, $\boldsymbol{P}$ and t. In the following, we introduce four kinds of generating functions depending on the choice of independent coordinates.

1. The generating function of the first kind $G \equiv G_1(\boldsymbol{q}, \boldsymbol{Q}, t)$ depends only on the old and new generalised coordinates. According to equation (2.45), we have

$$dG = \sum_{i=1}^{n} \left(\frac{\partial G_1}{\partial q_i} dq_i + \frac{\partial G_1}{\partial Q_i} dQ_i \right) + \frac{\partial G_1}{\partial t} dt = \sum_{i=1}^{n} p_i dq_i - \sum_{i=1}^{n} P_i dQ_i + (H' - H) dt.$$

   Comparing differential terms on both sides, we obtain the canonical transformation

$$p_i = \frac{\partial G_1}{\partial q_i}, \quad P_i = -\frac{\partial G_1}{\partial Q_i}, \quad H' = H + \frac{\partial G_1}{\partial t}. \tag{2.46}$$

2. The second generating function is obtained by reforming equation (2.45) with the Legendre transformation

$$d(G + \boldsymbol{Q} \cdot \boldsymbol{P}) = \boldsymbol{p} \cdot d\boldsymbol{q} + \boldsymbol{Q} \cdot d\boldsymbol{P} + (H' - H) dt$$

   The function $G_2(\boldsymbol{q}, \boldsymbol{P}, t) = G + \boldsymbol{Q} \cdot \boldsymbol{P}$, which depends on the old generalised coordinates and the new generalised momenta, is the generating function of the second kind. Comparing the partial differentials on both sides, we have another group of canonical transformation

$$p_i = \frac{\partial G_2}{\partial q_i}, \quad Q_i = \frac{\partial G_2}{\partial P_i}, \quad H' = H + \frac{\partial G_2}{\partial t}. \tag{2.47}$$

3. Following the same procedure of Legendre transformation, we introduce the generating function of the third kind $G_3(\boldsymbol{p}, \boldsymbol{Q}, t) = G - \boldsymbol{p} \cdot \boldsymbol{q}$ with the canonical transformation

$$q_i = -\frac{\partial G_3}{\partial p_i}, \quad P_i = -\frac{\partial G_3}{\partial Q_i}, \quad H' = H + \frac{\partial G_3}{\partial t}. \tag{2.48}$$

4. The generating function of the fourth kind is given by $G_4(\boldsymbol{p}, \boldsymbol{P}, t) = G + \boldsymbol{Q} \cdot \boldsymbol{P} - \boldsymbol{p} \cdot \boldsymbol{q}$ with the canonical transformation

$$q_i = -\frac{\partial G_4}{\partial p_i}, \quad Q_i = \frac{\partial G_4}{\partial P_i}, \quad H' = H + \frac{\partial G_4}{\partial t}. \tag{2.49}$$

The canonical transformations blur the distinguishment of generalised coordinates and generalised momenta. Since each new coordinate Q_i connects the old coordinates $\boldsymbol{q}$ and momenta $\boldsymbol{p}$, phase space can be viewed as a unified continuum just like spacetime continuum. We call the pair of dynamical variables $(\boldsymbol{q}, \boldsymbol{p})$ in Hamiltonian dynamics the *canonical conjugate variables.* The canonical transformation, which relates one pair of conjugate variables $(\boldsymbol{q}, \boldsymbol{p})$ to another conjugate pair $(\boldsymbol{Q}, \boldsymbol{P})$, preserves some important properties:

- *Symplectic condition*: $\boldsymbol{M}^T\boldsymbol{\Omega}\boldsymbol{M} = \boldsymbol{\Omega}$, where $\boldsymbol{\Omega}$ is the symplectic matrix defined in equation (2.34) and $\boldsymbol{M}$ is the Jacobian matrix of canonical transformation, i.e.,

$$\boldsymbol{M} \equiv \frac{\partial(\boldsymbol{Q}, \boldsymbol{P})}{\partial(\boldsymbol{q}, \boldsymbol{p})} = \begin{pmatrix} \partial \boldsymbol{Q}/\partial \boldsymbol{q} & \partial \boldsymbol{Q}/\partial \boldsymbol{p} \\ \partial \boldsymbol{P}/\partial \boldsymbol{q} & \partial \boldsymbol{P}/\partial \boldsymbol{p} \end{pmatrix}. \tag{2.50}$$

  Here, we use the convention $(\partial \boldsymbol{x}/\partial \boldsymbol{y})_{ij} \equiv \partial x_i/\partial y_j$ for two vectors $\boldsymbol{x}$ and $\boldsymbol{y}$.
- *Invariance of Poisson bracket:* $\{f, g\}_{\boldsymbol{q},\boldsymbol{p}} = \{f, g\}_{\boldsymbol{Q},\boldsymbol{P}}$, where f and g are two smooth functions and the Poisson brackets are defined in two coordinate systems. From identity (2.21), we directly have

$$\left\{Q_i, Q_j\right\}_{\boldsymbol{q},\boldsymbol{p}} = \left\{P_i, P_j\right\}_{\boldsymbol{q},\boldsymbol{p}} = 0, \quad \left\{Q_i, P_j\right\}_{\boldsymbol{q},\boldsymbol{p}} = \delta_{ij}. \tag{2.51}$$

Note that the phase flow given by equation (2.35), i.e.,

$$g_{t_0,t}: (\boldsymbol{q}(t_0), \boldsymbol{p}(t_0)) \to (\boldsymbol{q}(t_0 + t), \boldsymbol{p}(t_0 + t)) \equiv (\boldsymbol{Q}, \boldsymbol{P}) \tag{2.52}$$

is also a canonical transformation, as the canonical equations always hold during the time evolution of the Hamiltonian system. Because the canonical equations are $2n$ first-order differential equations, the trajectory in phase space is uniquely determined by the initial condition. Therefore, the inverse map of $g_{t_0,t}$ is also a canonical transformation, which maps each trajectory to a fixed point $(\boldsymbol{q}(t_0), \boldsymbol{p}(t_0))$.

Jacobian determinant For a square Jacobi matrix defined in the form of equation (2.50), we call the determinant of Jacobi matrix the *Jacobian determinant* or the *Jacobian.* One can prove the Jacobian has the following properties:

- $\left|\frac{\partial(X, Y)}{\partial(x, y)}\right| = -\left|\frac{\partial(X, Y)}{\partial(y, x)}\right| = -\left|\frac{\partial(Y, X)}{\partial(x, y)}\right|$
- $\left|\frac{\partial(X, y)}{\partial(x, y)}\right| = \left|\frac{\partial X}{\partial x}\right|_y$
- $\left|\frac{\partial(X, Y)}{\partial(x, y)}\right| = \left|\frac{\partial(X, Y)}{\partial(u, v)}\right| \cdot \left|\frac{\partial(u, v)}{\partial(x, y)}\right|.$

For the Jacobian of canonical transformation, we have

$$\left|\frac{\partial(\boldsymbol{Q}, \boldsymbol{P})}{\partial(\boldsymbol{q}, \boldsymbol{p})}\right| = \left|\frac{\partial(\boldsymbol{Q}, \boldsymbol{P})}{\partial(\boldsymbol{q}, \boldsymbol{P})}\right| \Big/ \left|\frac{\partial(\boldsymbol{q}, \boldsymbol{p})}{\partial(\boldsymbol{q}, \boldsymbol{P})}\right| = \left|\frac{\partial \boldsymbol{Q}}{\partial \boldsymbol{q}}\right|_{\boldsymbol{P}} \Big/ \left|\frac{\partial \boldsymbol{p}}{\partial \boldsymbol{P}}\right|_{\boldsymbol{q}} = 1. \tag{2.53}$$

Here, we have adopted the canonical transformation with the generating function of second kind and used the first two relations in equation (2.47) in the last step of equation (2.53). This means the canonical transformation preserves the phase space volume just like the Liouville's theorem. The canonical transformations also preserve the Poincaré–Cartan integral defined by equation (2.30).

2.1.4 Hamilton–Jacobi equation

The purpose of searching for canonical transformations is to get a simplified new Hamiltonian H'. The best canonical transformation is to make the new Hamiltonian $H' = 0$. As a result, the new canonical variables are all constants, i.e.,

$$Q_i = a_i, \quad P_i = b_i. \tag{2.54}$$

Here, a_i and b_i are $2n$ integral constants. This canonical transformation maps each trajectory in the old phase space to a fixed point $(\boldsymbol{a}, \boldsymbol{b})$ in the new phase space. This canonical transformation always exists, e.g., the inverse map of phase flow, as we discussed below equation (2.52). If we choose the generating function of the second kind $G_2(\boldsymbol{q}, \boldsymbol{b}, t)$ to make $H' = 0$, we have from the canonical transformation (2.47)

$$\boldsymbol{p} = \frac{\partial G_2}{\partial \boldsymbol{q}}, \quad \boldsymbol{a} = \frac{\partial G_2}{\partial \boldsymbol{b}} \tag{2.55}$$

and

$$\frac{\partial G_2}{\partial t} + H\left(\boldsymbol{q}, \frac{\partial G_2}{\partial \boldsymbol{q}}, t\right) = 0. \tag{2.56}$$

Since $\boldsymbol{b}$ is constant vector, we write the total time derivative of $G_2(\boldsymbol{q}, \boldsymbol{b}, t)$

$$\frac{dG_2}{dt} = \sum_{i=1}^{n} \frac{\partial G_2}{\partial q_i}\frac{dq_i}{dt} + \frac{\partial G_2}{\partial t} = \boldsymbol{p} \cdot \dot{\boldsymbol{q}} - H = L. \tag{2.57}$$

Therefore, the generating function G_2 is just the action given by equation (2.1)

$$S = \int_{t_1}^{t_2} L dt. \tag{2.58}$$

However, the action S defined here is an integral over a real physical path that satisfies Lagrange's equation [2] and should not be understood as a functional of $\boldsymbol{q}(t)$. For a fixed initial condition $\boldsymbol{q}(t_1)$, the action is a function of the final configuration $\boldsymbol{q}(t_2)$ and time $t = t_2$. Therefore, we can write equation (2.56) as the following

$$\frac{\partial S}{\partial t} + H\left(\boldsymbol{q}, \frac{\partial S}{\partial \boldsymbol{q}}, t\right) = 0, \tag{2.59}$$

which is called the *Hamilton–Jacobi equation*. Given the solution $S(\boldsymbol{q}, \boldsymbol{b}, t)$, we have $2n$ equations

$$\boldsymbol{p} = \frac{\partial S}{\partial \boldsymbol{q}}, \quad \boldsymbol{a} = \frac{\partial S}{\partial \boldsymbol{b}}, \tag{2.60}$$

that give the implicit relationships between $(\boldsymbol{q}, \boldsymbol{p})$ and $(\boldsymbol{a}, \boldsymbol{b})$. For $\det(\partial^2 S/\partial \boldsymbol{a}\partial \boldsymbol{b}) \neq 0$, the trajectory $(\boldsymbol{q}(t), \boldsymbol{p}(t))$ is the solution of the original Hamilton's equations with $(\boldsymbol{a}, \boldsymbol{b})$ determined by the initial condition. This statement is called the *Hamilton–Jacobi theorem*. Solving the Hamilton–Jacobi equation (2.59) is equivalent to solve Hamilton's equations.

Separation of variables

One important method to solve the Hamilton–Jacobi equation is the so-called *separation of variables* [2]. Suppose on the left-hand side of equation (2.59), one coordinate (e.g., q_1) and its derivative $\partial S/\partial q_1$ only appear in some function form $\varphi(q_1, \partial S/\partial q_1)$. In this case, we seek the action in the following form

$$S = S'(\boldsymbol{q}', t) + S_1(q_1), \tag{2.61}$$

where $S'(\boldsymbol{q}')$ represents the action without coordinate q_1, and the function $\varphi(q_1, \partial S/\partial q_1)$ is separated from the rest coordinates in equation (2.59). Therefore, the Hamilton–Jacobi equation becomes two separated equations

$$\varphi\left(q_1, \frac{dS_1}{dq_1}\right) = h_1 \tag{2.62}$$

$$\frac{\partial S'}{\partial t} + H\left(\boldsymbol{q}', \frac{\partial S'}{\partial \boldsymbol{q}'}, t\right) = 0, \tag{2.63}$$

where h_1 is a constant determined by initial condition. The first equation (2.62) is an ordinary differential equation, which is can be solved by the simple integral method. The number of independent variables in equation (2.63) has reduced by two. If we can separate all the n coordinates and time, the Hamilton–Jacobi equation is solved with the action in the form of

$$S = \sum_{i=1}^{n} W_i(q_i, \boldsymbol{h}) - E(\boldsymbol{h})t. \tag{2.64}$$

Here, $\boldsymbol{h} = (h_1, h_2, \ldots, h_n)$ is the set of integral constants. $E(\boldsymbol{h})$ is the energy constant as a function of $\boldsymbol{h}$ and can be obtained by plugging S back to the Hamilton–Jacob equation. Therefore, we can replace one element of $\boldsymbol{h}$ by the energy E, e.g., assigning $h_1 = E$.

2.2 Weakly-chaotic dynamics

2.2.1 Near-integrable systems

In section 2.1.4, we have discussed a dynamical system whose Hamilton–Jacobi equation can be completely solved by the method of separation of variables. In this case, the action without time part can be written in a sum form

$$W(\boldsymbol{q}, \boldsymbol{h}) \equiv \sum_{i=1}^{n} W_i(q_i, \boldsymbol{h}). \tag{2.65}$$

This is an example of completely *integrable system.* We define a new set of canonical momenta $\boldsymbol{I} = (I_1, I_2, \dots, I_n)$ by

$$I_i \equiv \frac{1}{2\pi} \oint_{\gamma_i} \frac{\partial}{\partial q_i} W_i(q_i, \boldsymbol{h}) dq_i, \tag{2.66}$$

where γ_i is a closed loop given by fixed integral constants $\boldsymbol{h}$. By replacing $\boldsymbol{h}$ by new momenta variables $\boldsymbol{I}$, the action can be rewritten by $W(\boldsymbol{q}, \boldsymbol{I}) = \sum_{i=1}^{n} \bar{W}_i(q_i, \boldsymbol{I})$. We then define a new set of canonical coordinates $\boldsymbol{\theta} = (\theta_1, \theta_2, \dots, \theta_n)$ by

$$\theta_i \equiv \frac{\partial W}{\partial I_i} = \sum_{i=1}^{n} \frac{\partial}{\partial I_i} \bar{W}(q_i, \boldsymbol{I}). \tag{2.67}$$

The new canonical momentum I_i is called the *action variable* and the new canonical coordinate θ_i is called the *angle variable*. For example, the orbit of a one dimensional (1D) harmonic oscillator $H_0(x, p) = \frac{1}{2}p^2 + \frac{1}{2}q^2$ is just the isoenergetic contour of $H_0(x, p) = h$ with enclosed elliptical area $\oint p \, \mathrm{d}x = 2\pi h$. The action variable is given by

$$I \equiv \frac{1}{2\pi} \oint p \, \mathrm{d}x = h. \tag{2.68}$$

The action $W(q, I)$ can be obtained from equations (2.60) and (2.64), i.e.,

$$W(q, I) = \int_{q_0}^{q} p dq = \int_{q_0}^{q} \sqrt{2I - q^2} dq \tag{2.69}$$

and the angle variable is given by

$$\theta = \frac{\partial W}{\partial I} = \int_{q_0}^{q} \frac{1}{\sqrt{2I - q^2}} dq. \tag{2.70}$$

By integrating the above equation and using $p = \partial W / \partial q$, we have the canonical transformation that connects the old and new phase space coordinates

$$q = \sqrt{2I} \sin(\theta + \theta_0), \quad p = \sqrt{2I} \cos(\theta + \theta_0) \tag{2.71}$$

with the initial value of angle variables $\theta_0 = \arcsin(q_0/\sqrt{2h})$.

2.2.1.1 Liouville integrability

The above discussion about integrable systems, which is solved by separated variables, can be extended to general Hamiltonian systems. The n dimensional Hamiltonian system is said to be *completely integrable* if there exists a set of n independent integrals of motions $(H_1, H_2, \dots, H_n)$ such that all the integrals of motion are in involution (see equation (2.25)) with each other, i.e.,

$$\{H_i, H_j\} = 0, \quad i, j = 1, 2, \dots, n. \tag{2.72}$$

Here, one integral of motion should take the Hamiltonian itself, e.g., $H_1 = H$. We define a manifold M_h as the set of phase space coordinates, which is determined by a set of n constant integrals of motion $\boldsymbol{h} = (h_1, h_2, \dots, h_n)$, i.e.,

$$M_h \equiv \{(\boldsymbol{q}, \boldsymbol{p}) | H_i(\boldsymbol{q}, \boldsymbol{p}) = h_i, \ i = 1, 2, \dots, n\}. \tag{2.73}$$

If there is only one integral of motion $h_1 = H$, M_h is the set of all the possible states (p, q) on the isoenergetic surface. Assuming all the integrals of motion are independent in the manifold M_h, the *Liouville's integrability theorem* states that [3, 5]

- M_h is a smooth manifold invariant under the phase flow with Hamiltonian $H = H_1$, that is, the trajectory starting from a point in M_h stays in the manifold M_h.
- If M_h is compact and connected, it is homeomorphic with an n dimensional torus

$$T^n \equiv \{(\theta_1, \theta_2, \dots, \theta_n), \bmod 2\pi\}. \tag{2.74}$$

- The phase flow in the manifold M_h is (quasi-)periodic, i.e.,

$$\frac{d}{dt}\boldsymbol{\theta} = \boldsymbol{\omega}(\boldsymbol{h}), \tag{2.75}$$

  where $\boldsymbol{\omega} = (\omega_1, \omega_2, \dots, \omega_n)$ is n-tuple of frequencies.
- The Hamilton–Jacobi equation with $H = H_1$ can be solved by finite algebraic operations and integrals of given functions.

It is also proved that there exists a set of variables that are functions of $\boldsymbol{h}$

$$\boldsymbol{I} \equiv (I_1, I_2, \dots, I_n) = \boldsymbol{I}(\boldsymbol{h}) \tag{2.76}$$

such that $\boldsymbol{I}$ and $\boldsymbol{\theta}$ are canonical variables satisfying Hamilton's equations

$$\frac{d\boldsymbol{I}}{dt} = -\frac{\partial H(\boldsymbol{I})}{\partial \boldsymbol{\theta}} = 0, \quad \frac{d\boldsymbol{\theta}}{dt} = \frac{\partial H(\boldsymbol{I})}{\partial \boldsymbol{I}} = \boldsymbol{\omega}(\boldsymbol{I}). \tag{2.77}$$

In the $(\boldsymbol{I}, \boldsymbol{\theta})$ action–angle coordinate system, the solution has the simple form

$$I(t) = I(0), \quad \theta(t) = \theta(0) + \omega(I(0))t. \tag{2.78}$$

It is crucial to find the canonical transformation $(q, p) \to (I, \theta)$ that makes the Hamiltonian only dependent of actions $H(I)$. The method given by equations (2.66) and (2.67) is an example to find such canonical transformation for systems with fully separable variables.

2.2.1.2 Tori and quasi-period motion

We discuss the characteristic properties of trajectories of a integrable system in phase space [3, 6–8]. The solution (2.78) shows the motion of integrable systems is regular in the action–angle coordinate system. For a set of fixed integrals of motion I, the corresponding angle variables $\theta = \{\theta_1, \theta_2, \dots, \theta_n\}$ change periodically with frequencies $\omega = (\omega_1, \omega_2, \dots, \omega_n)$ in an n dimensional torus T^n. We call the Hamiltonian system *non-resonant* if the components of frequency vector ω are independent, that is, for any nonzero integer vector $k = (k_1, k_2, \dots, k_n)$,

$$k \cdot \omega = \sum_{i=1}^{n} k_i\omega_i \neq 0, \quad k \in \mathbb{Z}^n \setminus \{0\}. \tag{2.79}$$

The motion of system is *quasi-periodic* when ω satisfies the non-resonance condition (2.79). In this case, any trajectory of system distributes uniformly and densely on the torus. In the infinitely-long time limit, the trajectory fills the whole torus. For a given dynamical variable $f(\theta, I) = f_0(I) + \sum_k f_k \exp(ik \cdot \theta)$, one can prove the average of $f(\theta, I)$ on the torus equals the long-time average, i.e.,

$$\frac{1}{(2\pi)^n}\int_0^{2\pi} f(\theta, I)d\theta = \lim_{T\to\infty}\frac{1}{T}\int_0^{T} f(\theta_0 + \omega t, I)dt = f_0(I). \tag{2.80}$$

This result is called *average theorem*. The integrable Hamiltonian system satisfying the average theorem is called *ergodic* [3].

The Hamiltonian system satisfies *resonance condition* if there exist a set of n nonzero integers $k^u = \{k_1^u, k_2^u, \dots, k_n^u\}$ such that

$$k^u \cdot \omega = \sum_{i=1}^{n} k_i^u\omega_i = 0, \quad k_i^u \in \mathbb{Z} \setminus \{0\}. \tag{2.81}$$

The *order of resonance* is given by $|k^u| \equiv \sum_{i=1}^{n}|k_i^u|$. Here, the superscript $u = 1, 2, \dots, \alpha$ represents the different combination of $\{k_i^u\}$ for a given resonance order $|k^u|$. For an n dimensional system, there are at most $\alpha = n - 1$ resonance conditions. In this case, we call the system is *completely resonant*. On the completely resonant tori, the trajectory of system is closed and thus *periodic*. We call the system is *partially resonant* if the number of resonance conditions is $1 \leqslant \alpha < n - 1$. In this case, the system has $n - \alpha$ independent frequencies. Thus, the motion of the system is periodic on the α dimensional torus and quasi-periodic in a lower-dimensional torus [3].

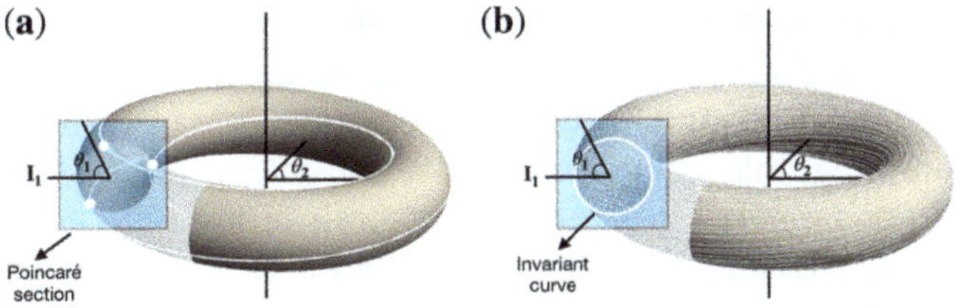

Figure 2.2. Motion of a 2D Hamiltonian system on the torus T^2. (a) Resonant motion with rational winding number $\nu = 1/3$. The Poincaré section on the lower-dimensional phase space I_1–θ_1 plane is composed of three discrete points. (b) Non-resonant motion with irrational winding number $\nu = (\sqrt{5} - 1)/2$. The quasi-period trajectory fills out the whole torus surface resulting in a continuous Poincaré section called an invariant curve.

In figure 2.2, we show the torus T^2 of a 2D system for a fixed (I_1, I_2). Suppose the frequencies along the θ_1 and θ_2 directions are ω_1 and ω_2 respectively. We call the ratio $\nu \equiv \omega_1/\omega_2$ the *winding number* or *rotation number*. When the winding number is a rational number $\nu = m/k$ with m and k two coprime integers, the trajectory is a periodic closed curve rotating m times along θ_1 direction and k times along θ_2 direction. When the winding number is an irrational number, the motion of system is quasi-periodic and the trajectory tiles the whole T^2 torus. A powerful tool to investigate the essential dynamics of high dimensional system from its myriad phase space trajectories is the *surface of section* or *Poincaré section*. Instead of looking at the continuous time evolution, the basic idea of Poincaré technique is to track the successive intersections of a trajectory or a set of trajectories with a chosen lower-dimensional 'plane' in phase space, which is usually a 2D plane spanned by a coordinate axis and the canonically conjugate momentum axis. The resultant surface of section is a set of interaction points on the phase space plane. We plot the Poincaré section of trajectories on the phase space plane (I_1, θ_1) by choosing fixed I_2 and θ_2. For the periodic motion, the Poincaré section of a single trajectory is a set of discrete points as shown in figure 2.2(a). For the quasi-period motion, a single trajectory fills out a curve on the surface of section, which is called the *invariant curve*, as shown in figure 2.2(b).

For a generic Hamiltonian system, there are an infinite number of tori. Some are resonant tori surrounded by periodic trajectories while others are non-resonant tori covered by quasi-periodic trajectories. Both resonant and non-resonant tori are infinite, and the number of non-resonant tori is much more than the number of resonant tori. This is similar to compare the rational numbers and irrational numbers. The measure of resonant tori is zero while the measure of non-resonant tori is 1. For n dimensional integral Hamiltonian system, the dimensions of phase space, isoenergetic manifold and torus are $2n$, $2n - 1$ and n respectively. For $n = 1$, the isoenergetic surface and the torus are the same 1D manifold. For $n = 2$, the 2D torus is embedded in the 3D isoenergetic manifold and divides the isoenergetic manifold into disconnected subregions. For $n > 2$, the torus, however, cannot divide the isoenergetic manifold into disconnected subregions any more. For this reason, the 2D system has different characteristic chaotic motion from that of systems with dimension $n \geqslant 3$, which will be discussed later in this section.

2.2.1.3 Small-divisor problem in near-integrable systems

Integrable systems have simple but rich periodic and quasi-periodic motions in phase space. However, there are quite a few integrable systems in reality. Most physical systems are non-integrable. For completely non-integrable systems that do not support any independent and in-involution integrals of motions, it is very difficult to study them analytically and should resort to numerical simulations. In order to reveal the role of non-integrability in dynamics, we consider the *near-integrable systems* whose Hamiltonian can be separated into the sum of a integrable part and a small perturbation,

$$H(\boldsymbol{I}, \boldsymbol{\theta}) = H_0(\boldsymbol{I}) + \epsilon H_1(\boldsymbol{I}, \boldsymbol{\theta}). \tag{2.82}$$

Our purpose is to find a canonical transformation from the old canonical variables $(\boldsymbol{I}, \boldsymbol{\theta})$ to the new canonical variables $(\boldsymbol{J}, \boldsymbol{\phi})$ such that the new transformed Hamiltonian only depends on the new momenta, i.e.,

$$H'(\boldsymbol{J}) = H(\boldsymbol{I}, \boldsymbol{\theta}). \tag{2.83}$$

Assuming the action (generating function) is $S(\boldsymbol{J}, \boldsymbol{\theta})$ and according to equation (2.60), we have

$$\boldsymbol{I} = \frac{\partial S(\boldsymbol{J}, \boldsymbol{\theta})}{\partial \boldsymbol{\theta}}, \quad \boldsymbol{\phi} = \frac{\partial S(\boldsymbol{J}, \boldsymbol{\theta})}{\partial \boldsymbol{J}}. \tag{2.84}$$

We seek the solution of $S(\boldsymbol{J}, \boldsymbol{\theta})$ in the form of series expansion to the first order

$$S(\boldsymbol{J}, \boldsymbol{\theta}) = \boldsymbol{J} \cdot \boldsymbol{\theta} + \epsilon S_1(\boldsymbol{J}, \boldsymbol{\theta}) + O(\epsilon^2), \tag{2.85}$$

which results in the following transformation

$$\boldsymbol{I} = \boldsymbol{J} + \epsilon \frac{\partial S_1}{\partial \boldsymbol{\theta}} + O(\epsilon^2), \quad \boldsymbol{\phi} = \boldsymbol{\theta} + \epsilon \frac{\partial S_1}{\partial \boldsymbol{J}} + O(\epsilon^2). \tag{2.86}$$

Plugging the above transformation to equation (2.83) and retaining the expansion to the first order of ϵ, we obtain

$$H_0(\boldsymbol{J}) + \epsilon \left[\boldsymbol{\omega} \cdot \frac{\partial S_1}{\partial \boldsymbol{\theta}} + H_1(\boldsymbol{J}, \boldsymbol{\theta}) \right] + O(\epsilon^2) = H'(\boldsymbol{J}) \tag{2.87}$$

with $\boldsymbol{\omega} = \partial H_0 / \partial \boldsymbol{J} = \partial H_0 / \partial \boldsymbol{I}$ (to the first order of ϵ). We write the perturbative Hamiltonian $H_1(\boldsymbol{I}, \boldsymbol{\theta})$ and action $S_1(\boldsymbol{I}, \boldsymbol{\theta})$ in the form of their Fourier expansions

$$H_1(\boldsymbol{I}, \boldsymbol{\theta}) = \sum_{k \neq 0} h_k(\boldsymbol{J}) e^{i\boldsymbol{k} \cdot \boldsymbol{\theta}}, \quad S_1(\boldsymbol{I}, \boldsymbol{\theta}) = \sum_{k \neq 0} s_k(\boldsymbol{J}) e^{i\boldsymbol{k} \cdot \boldsymbol{\theta}}. \tag{2.88}$$

Plugging them back to equation (2.87) and comparing the series expansion on both sides, we have the new Hamiltonian $H'(\boldsymbol{J}) = H_0(\boldsymbol{J})$ and the identities from Fourier components

$$s_k(\boldsymbol{J}) = -\frac{ih_k(\boldsymbol{J})}{\boldsymbol{k}\cdot\boldsymbol{\omega}} = -\frac{ih_k(\boldsymbol{J})}{k_1\omega_1 + k_2\omega_2 + \cdots + k_n\omega_n}. \tag{2.89}$$

Obviously, if all the possible winding numbers ω_i/ω_j $(i, j = 1, 2, \ldots, n)$ are rational, there always exist integers k_i $(i = 1, 2, \ldots, n)$ such that the denominator $\boldsymbol{k}\cdot\boldsymbol{\omega} = 0$ (resonance condition). Therefore, the first-order perturbative solution $S_1(\boldsymbol{J}, \boldsymbol{\theta})$ is divergent. Even for irrational winding numbers, the denominator $\boldsymbol{k}\cdot\boldsymbol{\omega}$ can be arbitrarily small as any irrational number can be approached by a series of rational numbers. This is the famous *small-divisor problem* in celestial mechanics [9]. The full solution of $S(\boldsymbol{J}, \boldsymbol{\theta})$ also depends on the higher-order expansions and the behaviour of numerator h_k compared to the denominator. Due to the zero or arbitrarily small denominator in the first-order expansion, one may ask if the resonant tori and non-resonant tori of an integrable system can survive under arbitrary small non-integrable perturbation. This problem was solved by the *Poincaré–Birkhoff theorem* and *Kolmogorov–Arnold–Moser theorem*, which will be discussed in the rest of this section.

2.2.2 Poincaré–Birkhoff theorem

In order to study the behaviour of periodic motion subjected to a perturbation, we introduce the concept of the *Poincaré map*, which plays an important role in the study of dynamical systems. In section 2.2.1, we have discussed the Poincaré section that is a set of intersection points of trajectories with a chosen phase space plane, e.g., the θ_1–I_1 phase plane for fixed (θ_2, I_2) as shown in figure 2.2. For autonomous Hamiltonian systems, a phase space trajectory is uniquely determined by a single point on the Poincaré section. Note that an n dimensional non-autonomous system can be reformed into a $n + 1$ autonomous system as discussed in section 2.1.2. The dynamical evolution can be viewed as a map that takes a point on the Poincaré surface and generates a new point on the Poincaré surface. This map is called the *Poincaré map*.

As an illustrative example, we take an $n = 2$ dimensional system as shown in figure 2.3. For a given rational winding number $\nu = \omega_1/\omega_2$, we take the Poincaré section by fixing the values of I_2 and θ_2. That is to say, we record the phase space coordinates (θ_1, I_1) stroboscopically every time interval $2\pi/\omega_2$. The resultant Poincaré map can be expressed by

$$T_0\begin{pmatrix}\theta_1\\ I_1\end{pmatrix} \equiv \begin{pmatrix}\theta_1 + 2\pi\nu(I_1)\\ I_1\end{pmatrix}. \tag{2.90}$$

Here, the winding number $\nu(I_1)$ is a function of I_1. The map T_0 is called the *twist map* [10, 11] if $\nu(I_1)$ is monotonic in a finite interval $I_- < I_1 < I_+$. In most physical systems, the twist map condition can be satisfied. We study the periodic motion with the rational winding number $\nu = m/k$ for a particular value of $I_1 = I^*$, where m and k are coprime integers. The Poincaré section of a single trajectory is a set of m discrete points on the θ_1–I_1 plane. These discrete points can be viewed as fixed points of a new map T_0^k, i.e.,

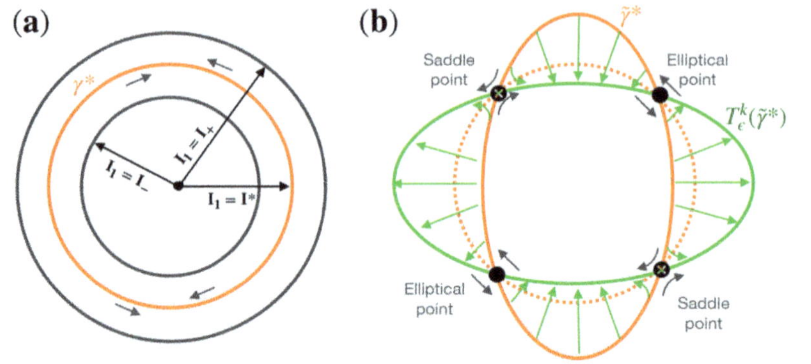

Figure 2.3. Poincaré–Birkhoff theorem. (a) Unperturbed Poincaré map T_0. γ^* (solid orange circle) represents the manifold with zero winding number of the k-fold Poincaré map T_0^k. (b) Perturbed Poincaré map T_ϵ. $\tilde{\gamma}^*$ (solid orange circle) represents the new manifold with zero winding number deformed from the old one (dashed orange circle). The cross points between $\tilde{\gamma}^*$ and $T_\epsilon^k(\tilde{\gamma}^*)$ (solid green circle) are the elliptical and saddled points of T_ϵ^k. Reproduced according to figure 1.8 in [3].

$$T_0^k\begin{pmatrix}\theta_1\\ I_1 = I^*\end{pmatrix} = \begin{pmatrix}\theta_1 + k2\pi\nu\\ I_1 = I^*\end{pmatrix} = \begin{pmatrix}\theta_1 + 2\pi m\\ I_1 = I^*\end{pmatrix} = \begin{pmatrix}\theta_1\\ I_1 = I^*\end{pmatrix}. \tag{2.91}$$

In other words, the effective winding number $\nu_k(I_1)$ is zero for the map T_0^k on the curve

$$\gamma^* = \{(\theta_1, I_1)|, I_1 = I^*\}. \tag{2.92}$$

Because T_0 and thus T_0^k are twist maps for $I_- < I_1 < I_+$, we assume for definiteness the effective winding number ν_k monotonically increases with I_1. As a result, the points (θ_1, I_1) within some interval $I^* < I_1 < I_+$ rotate anticlockwise while the points (θ_1, I_1) with $I_- < I_1 < I^*$ rotate clockwise as shown in figure 2.3(a).

Next, we consider a perturbation to the integrable Hamiltonian, i.e., $H = H_0(\boldsymbol{I}) + \epsilon H_1(\boldsymbol{I}, \boldsymbol{\theta})$. The corresponding Poincaré map is also modified as following

$$T_\epsilon\begin{pmatrix}\theta_1\\ I_1\end{pmatrix} \equiv \begin{pmatrix}\theta_1 + 2\pi\nu(I_1) + \epsilon f(\theta_1, I_1)\\ I_1 + \epsilon g(\theta_1, I_1)\end{pmatrix}, \tag{2.93}$$

where $f(\theta_1, I_1)$ and $g(\theta_1, I_1)$ are non-integrable perturbation functions depending on H_1. The explicit forms of f and g are irrelevant to our discussion. We only require that f and g are continuous functions. Most importantly, the map T_ϵ preserves phase space volume (area) for a Hamiltonian system. For sufficiently small perturbation parameter ϵ, T_ϵ maps the points to slightly different points compared to the map T_0. So we expect there is still some interval $\tilde{I}_- < I_1 < \tilde{I}_+$, where the winding number of T_ϵ^k is positive in some region close to $I_1 = \tilde{I}_+$ and negative in some region close to $I_1 = \tilde{I}_-$. By continuity, for every angle coordinate θ_1, there is a momentum $I_1^*(\theta_1) \in (\tilde{I}_-, \tilde{I}_+)$ such that the winding number of T_ϵ^k is zero, i.e., the angle θ_1 does not change under the map T_ϵ^k. Therefore, we can define a curve with zero winding number of T_ϵ^k, that is,

$$\tilde{\gamma}^* = \{(\theta_1, I_1)|, I_1 = I_1^*(\theta_1)\}. \tag{2.94}$$

Compared to the curve defined in equation (2.92), the new zero winding number curve (2.94) is no more a perfect circle but deformed in some way as shown in figure 2.3(b).

Note that the points on the curve $\tilde{\gamma}^*$ are generally not fixed points as the momentum I_1 can be changed by the map T_ϵ^k. The mapped curve $T_\epsilon^k(\tilde{\gamma}^*)$ has to cross with the curve $\tilde{\gamma}^*$ due to the area preservation of T_ϵ^k. Suppose the whole curve $T_\epsilon^k(\tilde{\gamma}^*)$ is surrounded by the circle $\tilde{\gamma}^*$, the phase space area enclosed by $T_\epsilon^k(\tilde{\gamma}^*)$ is not equal to (less than) the area enclosed by $\tilde{\gamma}^*$, and vice versa. Since the perturbation ϵH_1 is assumed to be sufficiently small, we exclude the situation that the mapped curve $T_\epsilon^k(\tilde{\gamma}^*)$ is totally separated from the curve $\tilde{\gamma}^*$. Because both $T_\epsilon^k(\tilde{\gamma}^*)$ and $\tilde{\gamma}^*$ are periodic in the θ_1 direction, they must have an even number of crossing points. These crossing points are truly fixed points as both coordinate θ_1 and momentum I_1 remain unchanged under the map T_ϵ^k. The fixed points can be either *elliptic points* or *saddle points.* The points near the new stable elliptical points will rotate around the elliptical point by the iteration of Poincaré map T_ϵ^k while the points near the new saddle points will go away along the unstable direction under the Poincaré map. From the phase flow near the crossing points as indicated in figure 2.3(b), half of them are stable elliptical points and other half are unstable saddle points. The results are summarised by the so-called *Poincaré–Birkhoff theorem* [12–15]:

For the perturbed twist map T_ϵ^k, there exist even number $2kq$ ($q = 1, 2, \ldots$) of fixed points on the closed curve generated by the Poincaré section of periodic trajectories of unperturbed integrable system. These fixed points are distributed on the curve alternatively as stable elliptic points and unstable saddle points.

Classification of fixed points Let T be a Poincaré map that is generated by the surface of section and $z^* = (\theta^*, J^*)^T$ is a fixed point on the Poincaré section. Suppose there is a point close to the fixed point, i.e., $z = z^*+\delta z$. After one iteration of Poincaré map, we calculate the deviation of mapped point to the fixed point

$$\overline{\delta z} \equiv T(z^*+\delta z) - z^* = \boldsymbol{M} \cdot \delta z + O(\delta z^2) \tag{2.95}$$

with the matrix $\boldsymbol{M} \equiv (\partial T/\partial z)|_{z=z^*}$. Assuming $\boldsymbol{\xi}$ is the eigenvector of $\boldsymbol{M}$, the solution of deviation is $\delta z = \rho\boldsymbol{\xi}$ with the eigenvalue ρ determined by

$$\det(\boldsymbol{M} - \rho \boldsymbol{I}) = 0. \tag{2.96}$$

We call the eigenvalue ρ the *characteristic multiplier*. As $\boldsymbol{M}$ is a real-valued matrix in our case, both ρ and ρ^* are the eigenvalues. For the Hamiltonian system, we have $\rho_1\rho_2 = 1$ since the Poincaré map should preserve the phase space area. Accordingly, we can classify the fixed point into three categories [1]:

- *Elliptic fixed points* for $\rho_1 = \rho_2^*$ and $|\rho_{1(2)}| = 1$. Elliptic points are linearly stable points in the sense that the points nearby trace ellipses around the fixed points.
- *Hyperbolic fixed points* or *saddle points* for positive real $\rho_1 > 1$ and $\rho_2 = \rho_1^{-1} < 1$. Saddle points are linearly unstable points in the sense that there is an exponentially expanding direction and an exponentially contracting direction. For negative multipliers, the fixed points are called *hyperbolic fixed points with reflection*.
- *Parabolic fixed points* for $\rho_1 = \rho_2 = \pm 1$. This happens at the critical (bifurcation) condition where the elliptic points become hyperbolic points or vice versa.

2.2.3 Kolmogorov–Arnold–Moser theorem

The Poincaré–Birkhoff theorem states that the resonant rational tori of an integrable system are unstable under the perturbation. Arbitrary small perturbation will destroy the resonant tori and split them into equal numbers of stable and unstable points. How about the non-resonant irrational tori? Are they also destroyed by arbitrary small perturbation? The unstability of resonant rational tori origins from the resonance among different real modes in physics. For a nonlinear integrable system, the resonant tori distribute densely in the whole isoenergetic manifold embedded in phase space. For this reason, Poincaré concluded that there is no independent integral of motion except the Hamiltonian itself for a near-integrable nonlinear system. This conclusion is correct for the *global* dynamics in phase space but does not answer the question of if there exist *local* integrals of motion. In 1954, Andrey Kolmogorov first answered this question and raised the so-called *invariant torus theorem* [16], which states that the non-resonant irrational tori of an integrable system can exist under sufficiently small perturbations. This statement was then extended in 1962 by Jürgen Moser for smooth twist maps using the Nash–Moser method [17]. Vladimir Arnold rigorously proved the invariant torus theorem in 1963 for an analytic perturbation Hamiltonian [9, 18]. This result is called the Kolmogorov–Arnold–Moser (KAM) theorem. It is not our purpose to provide a through mathematical proof of the KAM theorem in this book. We will explain the main conclusion of the KAM theorem and focus on the physical mechanism behind.

KAM theorem: suppose an n dimensional perturbed Hamiltonian written in the action–angle coordinate system

$$H(\boldsymbol{\theta}, \boldsymbol{I}) = H_0(\boldsymbol{I}) + \epsilon H_1(\boldsymbol{\theta}, \boldsymbol{I}), \tag{2.97}$$

where $H_0(\boldsymbol{I})$ is the unperturbed integrable Hamiltonian and $H_1(\boldsymbol{\theta}, \boldsymbol{I})$ is a non-integrable perturbation. For a given torus of integrable system $\boldsymbol{I} = \boldsymbol{I}_0$, there are conditions as follows.

1. *Analytical condition*: the Hamiltonian $H(\boldsymbol{\theta}, \boldsymbol{I})$ by analytic continuation is a real-valued analytic function in the region $\Sigma_0 = \{(\boldsymbol{\theta}, \boldsymbol{I}) || \mathrm{Im}(\boldsymbol{\theta})| \leqslant \rho, |\boldsymbol{I}_0 - \boldsymbol{I}| \leqslant s\}$.

2. *Strong non-resonant condition*: there exist $D = D(\boldsymbol{\omega}) > 0$ and $\mu > n - 1$ such that, for any nonzero integer vector $\boldsymbol{k} = (k_1, k_2, \ldots, k_n)$,

$$|\boldsymbol{k} \cdot \boldsymbol{\omega}| \equiv |\sum_{i=1}^{n} k_i\omega_i| \geqslant D(\boldsymbol{\omega})K^{-\mu}, \quad \text{where} \quad K \equiv \sum_{i=1}^{n} |k_i|. \tag{2.98}$$

3. *Non-degenerate condition*: the unperturbed system satisfies

$$\alpha \equiv \det\left[\frac{\partial \boldsymbol{\omega}(\boldsymbol{I})}{\partial \boldsymbol{I}}\right] = \det\left[\frac{\partial^2 H_0(\boldsymbol{I})}{\partial \boldsymbol{I}^2}\right] \neq 0. \tag{2.99}$$

4. *Bounded condition*: $\forall \epsilon > 0$, $\exists\, d = d(\epsilon, D, \mu, s, \rho) > 0$ such that $|H| < d$ in Σ_0.

If the system satisfies all the above four conditions, the Hamiltonian (2.97) has an n dimensional invariant torus

$$\boldsymbol{I} = \boldsymbol{I}_0 + \boldsymbol{\nu}(\boldsymbol{\xi}, \epsilon), \quad \boldsymbol{\theta} = \boldsymbol{\xi} + \boldsymbol{\mu}(\boldsymbol{\xi}, \epsilon), \quad \text{with} \quad \frac{d\boldsymbol{\xi}}{dt} = \left(\frac{\partial H}{\partial \boldsymbol{I}}\right)\bigg|_{\boldsymbol{I} = \boldsymbol{I}_0}. \tag{2.100}$$

Here, $\boldsymbol{\nu}$, $\boldsymbol{\mu}$ are 2π-periodic real-valued analytic function defined in the complex field $|\mathrm{Im}(\boldsymbol{\xi})| \leqslant \rho/2$, and the invariant torus (2.100) is sufficiently close to the corresponding invariant torus of unperturbed system, i.e., $|\boldsymbol{\nu}| + |\boldsymbol{\mu}| < s$. The first and second conditions guarantee that the series expansion in equation (2.88), i.e., $S_1(\boldsymbol{I}, \boldsymbol{\theta}) = \sum_{\boldsymbol{k}\neq 0} s_{\boldsymbol{k}}(\boldsymbol{J}) e^{i\boldsymbol{k}\cdot\boldsymbol{\theta}}$, converges nearly for any non-resonant frequency $\boldsymbol{\omega}$. To explain this, we write $H_1(\boldsymbol{I}, \boldsymbol{\theta}) = \sum_{\boldsymbol{k}\neq 0} h_{\boldsymbol{k}} e^{i\boldsymbol{k}\cdot\boldsymbol{\theta}}$ and calculate the Fourier coefficients

$$h_{\boldsymbol{k}} = \frac{1}{(2\pi)^n}\int_0^{2\pi} d\boldsymbol{\theta} H_1(\boldsymbol{I}, \boldsymbol{\theta}) e^{-i\boldsymbol{k}\cdot\boldsymbol{\theta}} = \frac{1}{(2\pi)^n}\int_{0-i\rho}^{2\pi-i\rho} d\boldsymbol{\theta} H_1(\boldsymbol{I}, \boldsymbol{\theta}) e^{-i\boldsymbol{k}\cdot\boldsymbol{\theta}}. \tag{2.101}$$

Here, the second equality comes from the first analytic condition of H_1 in the complex $\boldsymbol{\theta}$ plane. Therefore, there exist $B(\boldsymbol{\omega}) > 0$ such that $|h_{\boldsymbol{k}}| \leqslant B(\boldsymbol{\omega}) e^{-K\rho}$, where the $\boldsymbol{\omega}$ dependence of B can be understood from the average theorem (2.80). This inequality means the amplitude of $h_{\boldsymbol{k}}$ decreases exponentially with $K = \sum_{i=1}^{n} |k_i|$. Furthermore, according to the second condition of strong resonance, the denominator of $s_{\boldsymbol{k}}(\boldsymbol{J}) = -ih_{\boldsymbol{k}}/\boldsymbol{k}\cdot\boldsymbol{\omega}$ decreases as a function of K no more than power law, i.e., $|\boldsymbol{k}\cdot\boldsymbol{\omega}| \geqslant D(\boldsymbol{\omega})K^{-\mu}$. Therefore, the sum of first-order expansion in S_1 converges. In the same spirit, the KAM theorem solved the small-divisor problem in the high-order expansion of ϵ. Note that not all the non-resonant frequencies $\boldsymbol{\omega}$ satisfy the strong resonant condition. However, the *irrational condition theorem* [3, 19] guarantees that the measure of the set of frequencies that violate the strong non-degenerate condition is zero. Therefore, there are indeed an infinite invariant tori of integrable system which are destroyed by perturbation, but the probability of finding them approaches zero as $\epsilon \to 0$ [19].

The third condition of non-degeneracy essentially means the system is nonlinear as the frequency $\boldsymbol{\omega}$ changes with action variables $\boldsymbol{I}$. For the systems discussed by the KAM theorem, the nonlinearity plays a crucial role of stabilising the invariant tori, and the perturbation parameter ϵ should not be too large to overcome the nonlinearity. It would be more acceptable to write down the condition of applicability of the KAM theorem as [20]

$$\epsilon < C\alpha^{\delta}, \quad (\epsilon, \alpha \to 0), \tag{2.102}$$

where $\delta > 0$ and C is a constant independent (or slightly dependent) on the nonlinearity α. In some cases, the condition (2.102) can be violated but the invariant tori still persists under small perturbation, e.g., the harmonic oscillator with $\alpha = 0$ subjected to weak periodic driving field. We will discuss the situation beyond KAM theory in section 2.2.5.

2.2.4 Chaotic motion

The Poincaré–Birkhoff theorem states that, under an arbitrarily small perturbed twist map T_ϵ^k, the surface of the section of a resonant torus splits into an equal number of elliptic points and saddle points. In this section, we will study in detail the perturbed behaviour near the saddle points and the resultant chaotic motion [1, 3].

A saddle point has both stable and unstable directions, which can be obtained by solving the eigenequation (2.96) of the linearised Poincaré map. We define the *stable manifold* W^s of a single saddle point as the set of points that are asymptotic to the saddle point *forward* in time by the map T_ϵ^k, and the *unstable manifold* W^u as the set of points that are asymptotic to the saddle point *backward* in time by the map $(T_\epsilon^k)^{-1}$. In figure 2.4, the unstable and stable manifolds are represented by red and blue curves respectively. We emphasise that the (un)stable manifold is not a single trajectory of system but the collection of intersection points on the Poincaré section generated by many trajectories. For an integrable system, the stable manifold W^s and unstable manifold W^u coincide and connect saddle points. In this case, they are called the *separatrices*. The points on the two different sides of a separatrix separate exponentially in opposite directions when they approach the saddle point forward or backward in time.

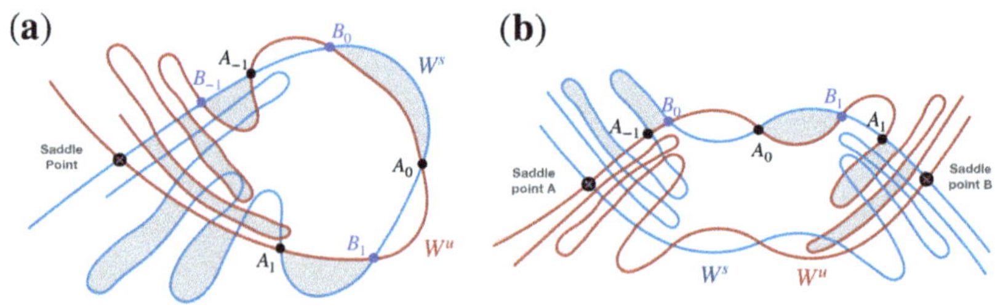

Figure 2.4. (a) Homoclinic intersection of a single saddle point. (b) Heteroclinic intersection of two saddle points. The red (blue) curves represent unstable (stable) manifolds of corresponding saddle points with crossing points labelled by A_i and B_i. The shaded regions preserve phase space area according to Liouville's theorem. Reproduced according to figure 1.11 in [3].

For non-integrable systems, the stable and unstable manifolds do not coincide in general. Instead, they will cross due to the area preservation and the bounded domain in phase space. The crossing is called *homoclinic intersection* if W^u and W^s belong to the same saddle point and the crossing point is called *homoclinic point* as shown in figure 2.4(a). The crossing is called *heteroclinic intersection* if W^u and W^s belong to different saddle points and the crossing point is called *heteroclinic point* as shown in figure 2.4(b). Note that the W^s or W^u cannot cross themselves. If $W^{s(u)}$ crosses itself, there will be a loop and the loop is mapped forward (backward) arbitrarily close to the saddle point. This contradicts the fact that the linearised Poincaré map of the saddle point does not have loops [1]. If manifolds W^s and W^u cross each other once, e.g., the interaction point A_0 shown in figure 2.4, there are an infinite number of intersection points generated by the map T_ϵ^k both forward and backward in time, which are labelled by $A_{\pm 1}$, $A_{\pm 2}$, … in the same plot. In fact, there are more intersection points of W^s and W^u. If we consider little segments of W^s and W^u near the intersection point A_0, the mapped segments near the next intersection point A_1 must keep both its area and orientation (the sign of area) invariant, see equation (2.41), according to Liouville's theorem. The preservation of orientation implies there is at least one more intersection point (e.g., B_0) between the successive crossing points (e.g., A_0 and A_1).

On the one hand, as the two neighbouring intersection points, e.g., A_0 and B_0, approach the saddle points, the distance between them get closer and closer. On the other hand, the area enclosed by the lobe between A_0 and B_0 keeps invariant according to Liouville's theorem as shown by the shaded regions in figure 2.4. This means the lobe must stretch in other direction to preserve its area, and thus the length of lobe in the stretched direction grows exponentially (maybe not exactly). The above discussion holds for all the other lobes enclosed by intersection points. Because the accessible domain in phase space is bounded in general, the lobes will wrap in the region near separatrix and form tendrils near the saddle points. The tendrils of W^s and W^u can cross again and create new homoclinic or heteroclinic points. The new intersection points repeat the complex behaviours near and saddle points and the separatrix region. The process goes on and on. The resultant complex dynamics near the saddle points and the separatrix region becomes *chaotic* in the end.

2.2.4.1 Stochastic layer and Arnold diffusion

If the non-integrable perturbation is weak, the regions of chaotic motion are separated by KAM tori (invariant curves) and form the so-called *chaotic layer* or *stochastic layer* as shown in figure 2.5(a). It is called *local chaos* if the width of the stochastic layer is thin. As the non-integrable perturbation increases, more and more KAM invariant curves are destroyed and the width of stochastic layer becomes thicker and thicker. When the perturbation becomes stronger, some separated chaotic regions can connect each other and form a *stochastic web* penetrating the phase space. If the perturbation is strong enough beyond a critical point, all the chaotic regions are connected, which is called *global chaos*. The regular regions of undestroyed KAM tori around the elliptic points look like

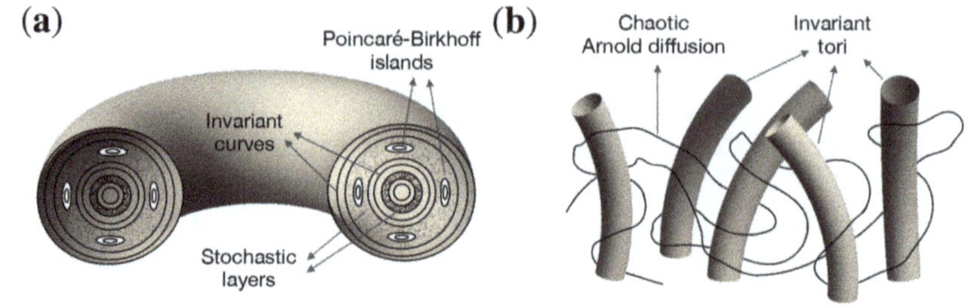

Figure 2.5. (a) Stochastic layers and Poincaré–Birkhoff islands for an $n = 2$ dimensional KAM system. (b) Arnold diffusion among invariant tori for an $n > 2$ dimensional KAM system. Reproduced according to figure 1.20 in [3].

isolated islands in the sea of chaos, which are called the *Poincaré–Birkhoff islands* or the *Birkhoff islands*. The Birkhoff islands may form a regular pattern in phase space, i.e., a chain of Birkhoff islands as shown in figure 2.5(a).

The system can diffuse inside the stochastic layer. For an n dimensional integrable system $H = H_0(\boldsymbol{J})$, the motion of system is constrained on an n dimensional torus embedded in $2n$ phase space. If the system is perturbed, the chaotic motion happens near the intersection region (stochastic layer) of the isoenergetic manifold $H = H_0(\boldsymbol{J}) + \epsilon H_1(\boldsymbol{J}, \boldsymbol{\theta}) = E$ and the resonant manifold determined by $\boldsymbol{k} \cdot \boldsymbol{\omega}(\boldsymbol{J}) = 0,\ \boldsymbol{k} \in \mathbb{Z}^n \equiv \{0\}$. The isoenergetic manifold and the resonant manifold are both $2n - 1$ dimensional in general. As a result, the intersection of them is $2(n - 1)$ dimensional. For the near-integrable system, there exist n dimensional KAM tori. For an $n = 2$ dimensional system, its $n = 2$ dimensional KAM tori can divide the $2n - 1 = 3$ dimensional isoenergetic manifold into isolated $2(n - 1) = 2$ dimensional resonant manifolds together with its neighbourhood, which are $2 + 1 = 3$ dimensional. Therefore, the chaotic motion in the stochastic layers cannot cross the invariant curves as shown in figure 2.5(a). However, for an $n = 3$ dimensional system, its $n = 3$ dimensional KAM tori cannot divide the $2n - 1 = 5$ dimensional isoenergetic manifold into isolated $2(n - 1) = 4$ dimensional resonant manifolds together with its neighbourhood, which are $4 + 1 = 5$ dimensional. This analysis is also valid for $n \geqslant 3$ dimensional system where the KAM tori cannot divide the isoenergetic manifold into isolated subregions and restrain the diffusion of chaotic motion in the stochastic layers. In this case, the intersections of all the possible resonant manifolds with the isoenergetic manifold can connect and form the so-called *Arnold web*, which is the first example of stochastic web proposed by Arnold in 1964 [21, 22]. From one point on the web, the motion can randomly reach any point on the web, which is called *Arnold diffusion* [23], see figure 2.5(b).

2.2.5 Non-KAM dynamics

For the systems that satisfy the KAM theorem, the nonlinearity plays an important role to stabilise the system. When the resonant tori are destroyed by small perturbation, the nonlinearity leads the system to escape from the resonance condition. Let us emphasise that the applicability of KAM theory needs a finite nonlinear parameter of the unperturbed system $\alpha = \det(\partial^2 H_0/\partial \boldsymbol{I}^2)$ and a fairly weak perturbation $\epsilon < C\alpha^\delta$, see equation (2.102).

In this section, we will go beyond KAM theory and consider the opposite condition that there is a finite perturbation ($\epsilon \neq 0$) but the unperturbed system has no nonlinearity ($\alpha = 0$). Without the protection of nonlinearity, will the regular dynamics of a linear system survive under the perturbation? If some regular motions still exist, how about the size of chaotic regions (stochastic layers) in phase space? The general answer is regular motions can survive under sufficiently small perturbation but the separatrices are destroyed and replaced by the stochastic web whose width tends to be zero as perturbation $\epsilon \to 0$ [20].

2.2.5.1 Kicked harmonic oscillator

We consider a concrete example of dynamical system with linear unperturbed Hamiltonian, i.e., the kicked harmonic oscillator (KHO) [20]:

$$H = \frac{1}{2}(p^2 + \omega_0^2 x^2) - \frac{\omega_0}{T} K \cos x \sum_{n\in\mathbb{Z}} \delta\left(\frac{t}{T} - n\right). \tag{2.103}$$

The unperturbed system ($K = 0$) is a linear oscillator with characteristic frequency ω_0 and unit mass. The perturbation is a periodically kicked cosine lattice with kicking amplitude K and period T. For the sake of simplicity, we take the resonant condition

$$q = 2\pi\omega_0^{-1}/T \in \mathbb{Z}. \tag{2.104}$$

Here, q represents the number of kicks during one harmonic period $2\pi/\omega_0$. By setting $p \to p/\omega_0$, $\tau \to t/T$ and $H \to HT/\omega_0$, we rewrite the Hamiltonian in the dimensionless form of

$$H(x, p, t) = \frac{\omega_0 T}{2}(p^2 + x^2) - K \cos x \sum_{n\in\mathbb{Z}} \delta(\tau - n). \tag{2.105}$$

The canonical equations are given by

$$\frac{dp}{d\tau} = -\frac{\partial H}{\partial x} = -\frac{2\pi}{q}x + K \sin x \sum_{n\in\mathbb{Z}} \delta(\tau - n), \qquad \frac{dx}{d\tau} = \frac{\partial H}{\partial p} = \frac{2\pi}{q}p. \tag{2.106}$$

We label the values of phase space coordinates just after the nth kick by $x_n \equiv x(n + 0^+)$ and $p_n \equiv p(n + 0^+)$. By integrating the above equations, we obtain the following discrete map

$$T_q: \begin{array}{l} x_{n+1} = x_n \cos\dfrac{2\pi}{q} + (p_n + K \sin x_n) \sin\dfrac{2\pi}{q} \\[2ex] p_{n+1} = -\, x_n \sin\dfrac{2\pi}{q} + (p_n + K \sin x_n) \cos\dfrac{2\pi}{q} \end{array}. \tag{2.107}$$

The discrete map T_q can be viewed as the Poincaré map of the Hamiltonian (2.103) by the procedure of autonomizing a time-dependent system (section 2.1.2.2) and recording the dynamics at stroboscopic time instants $\tau = n + 0^+$ with $n \in \mathbb{Z}$.

In figure 2.6 and figure 2.7, we show the Poincaré sections generated by the map T_q for several integers q from the randomly sampled points in phase space. We summarise some important properties as follows.

- Chaotic motion penetrates the whole phase space forming a stochastic web for arbitrarily small K. This a crucial difference from the weak chaotic motion for $n \leqslant 2$ dimensional system predicted by KAM theorem.
- Invariant curves and regular motions exist inside the meshes of stochastic web. The size of web meshes does not change for finite K but will disappear for $K = 0$. The width of stochastic web increases as K increases.
- Stochastic webs exhibit crystalline symmetries for $q \in \{3, 4, 6\}$. The web for $q = 4$ has square lattice symmetry while the webs for $q \in \{3, 6\}$ have hexagonal lattice symmetry. The motions of system for $q \in \{1, 2\}$ are trivial and regular.
- Stochastic webs for $q = 5$ and $q > 6$ have no lattice symmetries but quasicrystal symmetries. It was conjectured that unlimited quasicrystal type stochastic webs exist for arbitrarily small K [20].

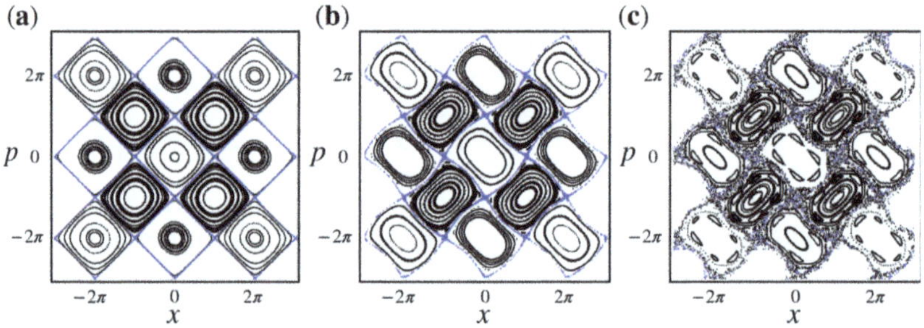

Figure 2.6. Poincaré section of kicked harmonic oscillator. Parameters: $q = 4$, $K = 0.1$ (a), $q = 4$, $K = 1.0$ (b), and $q = 4$, $K = 1.5$ (c). The initial states are randomly sampled in phase space. The set of blue dots is the single chaotic trajectory that starts from the saddle point $(\pi, 0)$ and diffuses over the whole stochastic layer.

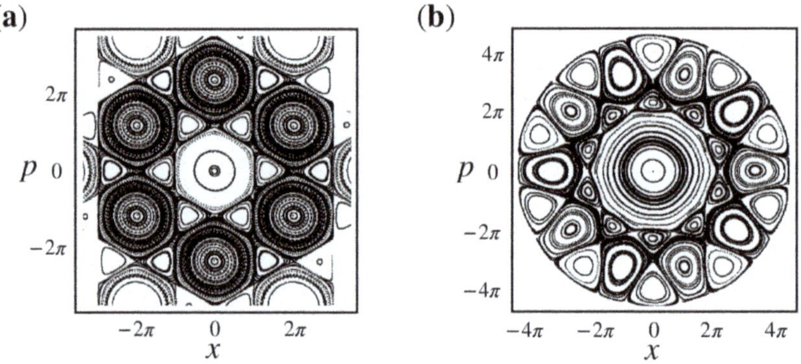

Figure 2.7. Poincaré section of kicked harmonic oscillator. Parameters: $q = 3, 6$, $K = 0.1$ (a) and $q = 5$, $K = 0.1$ (b). The skeleton of stochastic layer for $q = 3, 6$ exhibits a hexagonal lattice structure while the stochastic web skeleton for $q = 5$ has 5-fold quasicrystal symmetry.

2.2.5.2 Stochastic web skeleton

To understand the complex pattern of KHO in the Poincaré section, we introduce the polar coordinates (ρ, ϕ) via $x = \rho\cos\phi$ and $p = \rho\sin\phi$. Note that the polar coordinates (ρ, ϕ) are not canonical coordinates. The canonical coordinates are $(I = \rho^2/2, \phi)$. We transform into a rotating frame with frequency $2\pi/q$ via the generating function of the second kind

$$G_2 = \left(\phi - \frac{2\pi}{q}\tau\right)\tilde{I}.$$

Here, $\tilde{I} \equiv \tilde{\rho}^2/2$ is the new momentum, and $\tilde{\phi} = \partial G_2/\partial\tilde{I} = \phi - 2\pi\tau/q$ is the new angle variable from equation (2.47). For simplicity, we neglect the tildes and write the transformed Hamiltonian as follows

$$\tilde{H} = H + \frac{\partial G_2}{\partial \tau} = -K\cos[\rho\cos(\phi + 2\pi\tau/q)]\sum_{n=-\infty}^{+\infty}\delta(\tau - n). \tag{2.108}$$

Note that the unperturbed harmonic term disappears in the new Hamiltonian. Consider the following identities [20]

$$\begin{aligned}\sum_{n=-\infty}^{+\infty}\delta(\tau - n) &= \sum_{j=1}^{q}\sum_{m=-\infty}^{+\infty}\delta(\tau - (mq + j)),\\ \sum_{m=-\infty}^{+\infty}\delta(\tau - j - mq) &= \frac{1}{q}\sum_{m=-\infty}^{+\infty}\exp\left(i2\pi m\frac{\tau - j}{q}\right).\end{aligned} \tag{2.109}$$

We regroup the terms in the transformed Hamiltonian as follows

$$\tilde{H} = H_q + V_q \tag{2.110}$$

with

$$\begin{aligned}H_q &= -\frac{1}{2}\Omega_q\sum_{j=1}^{q}\cos\xi_j,\\ V_q &= -\Omega_q\sum_{j=1}^{q}\cos\xi_j\sum_{m=1}^{q}\cos\left(\frac{2\pi m}{q}(\tau - j)\right)\end{aligned} \tag{2.111}$$

where

$$\Omega_q \equiv \frac{2K}{q},\quad \xi_j \equiv \rho\cos\left(\phi + \frac{2\pi}{q}j\right) = x\cos\left(\frac{2\pi}{q}j\right) - p\sin\left(\frac{2\pi}{q}j\right). \tag{2.112}$$

The stationary part H_q is called the *skeleton Hamiltonian* [20] that defines the skeleton of the stochastic web. All saddle points of skeleton Hamiltonian $H_q(x, p)$ have the same energy E_s. For $q \in \{3, 4, 6\}$, all the separatrices that connect the saddle points are isocontours of $H(x_s, p_s) = E_s$. The skeletons of webs are unbounded and tile the whole phase space plane with square or hexagonal lattice symmetry. For $q = 5$ and $q > 6$, the skeleton Hamiltonian has the same quasicrystal symmetry as the stochastic web. However, the saddle points no longer belong to some unique value of energy and they distribute in a complicated manner with energy [20].

2.2.5.3 *Width of stochastic layer*

The width of stochastic web can be calculated using the Hamiltonian in the rotating frame. We take the case of $q = 4$ for example and write the Hamiltonian

$$\tilde{H} \approx H_4 + V_4 \tag{2.113}$$

with

$$\begin{aligned} H_4 &= -\,\Omega_4(\cos x + \cos p) \\ V_4 &= -\,2\Omega_4(\cos x - \cos p)\cos(\pi\tau) + \cdots, \end{aligned} \tag{2.114}$$

where the dots in V_4 include the $m \geqslant 3$ terms. Note that odd m terms give zero contribution to V_4. For the unperturbed Hamiltonian H_4, the separatrices are defined by the following equations with $H_4 = 0$, i.e., $x_s = \pm(p_s + \pi) + 2\pi k$ with $k \in \mathbb{Z}$. From the canonical equations, we derive the motion on the separatrices

$$\dot{x}_s = \Omega_4 \sin p_s = \pm\Omega_4 \sin x_s \tag{2.115}$$

$$\dot{p}_s = -\Omega_4 \sin x_s = \mp\Omega_4 \sin x_s. \tag{2.116}$$

The solution is

$$\sin p_s = -\sin x_s = -\frac{1}{\cosh[\Omega_4(\tau - \tau_n)]} \tag{2.117}$$

where τ_n is the point of reference in time. Using the Poisson bracket (2.23), we calculate the change of H_4 near the separatrices under the perturbation V_4,

$$\frac{dH_4}{dt} = \{H_4, \tilde{H}\} = -4\Omega_4^2 \sin x_s \sin p_s \cos(\pi\tau). \tag{2.118}$$

From equations (2.117) and (2.118), we can obtain the change of H_4 energy in the vicinity of the separatrix

$$\Delta H_4 = 4\Omega_4^2 \int_{\tau_n-\infty}^{\tau_n+\infty} d\tau \frac{\cos \pi\tau}{\cosh^2[\Omega_4(\tau - \tau_n)]} = 4\pi^2 \frac{\cos \pi\tau_n}{\sinh^2(\pi^2/2\Omega_4)}. \tag{2.119}$$

For $\Omega_4 = K/2 \ll 1$, we can get a rough estimation of the width of stochastic layer

$$\Delta H_4 \approx 8\pi^2 \cos(\pi\tau_n) \exp\left(-\frac{\pi^2}{K}\right). \tag{2.120}$$

This estimation already shows the main feature that the stochastic layer width is exponentially suppressed by the small perturbation K. A refined result omitting the reference time is given in [24]

$$\Delta H_4 = \frac{32\pi^2}{K}\exp\left(-\frac{\pi^2}{K}\right). \tag{2.121}$$

The neglected terms in V_4 give a correction of the order $\exp(-2\pi^2/K)$. The above analysis can be applied to other integers of q. In general, the width of stochastic web is approximately $\Delta H_q \sim \exp(-C_q/K)$ where the constant $C_q > 0$ increases somewhat with the growth of q [24].

The stochastic web divides phase space and imposes a symmetry by the web symmetry. The regular motions (Birkhoff islands) enclosed in the web meshes form a lattice in phase space, which we call the *phase space lattices*. As long as the width of web is thin enough, the phase space lattice dominates the dynamics of system. In fact, our book focuses on the regular dynamics in phase space lattice instead of the chaotic motions in the stochastic web.

2.3 Quantum mechanics

In this section, we discuss the quantum dynamics of the periodically driven (Floquet) system, which is governed by the Schrödinger equation

$$i\hbar\frac{\partial}{\partial t}|\psi(x, t)\rangle = H(x, t)|\psi(x, t)\rangle, \tag{2.122}$$

where the Hamiltonian is a periodic function of time $H(t) = H(t + T)$. The Schrödinger equation is closely related to but cannot be derived from the classical Hamilton–Jacobi equation (2.59). Essentially, the classical Hamilton–Jacobi equation (2.59) is in general nonlinear with respect to the action S while the Schrödinger equation is linear with respect to the quantum state $|\psi(x, t)\rangle$. The general framework to treat periodically driven systems in quantum mechanics is called *Floquet theory*. We will introduce the basics of Floquet theory in section 2.3.1 and the technique of Floquet–Magnus expansion in section 2.3.2.

2.3.1 Floquet theory

The discrete time translation symmetry of Floquet systems, i.e., $H(t) = H(t + T)$, enables us to use the Floquet formalism [25–27]. We first rewrite the Schrödinger equation (2.122) in an alternative form ($\hbar$=1)

$$\mathcal{H}(x, t)|\psi(x, t)\rangle = 0 \quad \text{with} \quad \mathcal{H}(x, t) \equiv H(x, t) - i\frac{\partial}{\partial t}, \tag{2.123}$$

where x represents the coordinates of particles or Pauli matrices of spins. The time periodicity of $\mathcal{H}(x, t) = \mathcal{H}(x, t + T)$, similar to the Bloch theorem in solid state physics, enforces the eigensolutions of equation (2.123) in the form of Floquet solution [25–29]

$$|\psi_\alpha(x, t)\rangle = e^{-i\epsilon_\alpha t}|\varphi_\alpha(x, t)\rangle \quad \text{with} \quad |\varphi_\alpha(x, t)\rangle = |\varphi_\alpha(x, t + T)\rangle. \tag{2.124}$$

Here, $|\varphi_\alpha(x, t)\rangle$ is called the *Floquet mode* and ϵ_α is called the *quasienergy*. The terminology quasienergy is analogous to the quasimomentum characterising the Bloch eigenstates in a crystalline solid. By substituting equation (2.124) into equation (2.123), we find

$$\mathcal{H}(x, t)|\varphi_\alpha(x, t)\rangle = \epsilon_\alpha|\varphi_\alpha(x, t)\rangle. \tag{2.125}$$

According to equation (2.124), the new state $|\varphi_{\alpha,n}(x, t)\rangle \equiv e^{in\Omega t}|\varphi_\alpha(x, t)\rangle$ $(n = 0, \pm 1, \pm 2, \ldots)$ is also the Floquet mode but with the shifted quasienergy $\epsilon_{\alpha,n} \equiv \epsilon_\alpha + n\Omega$. Therefore, we can map all the quasienergies into a first Brillouin zone, e.g., $0 \leqslant \epsilon_\alpha < \Omega$ that is a real-valued variable modulo $\Omega = 2\pi/T$.

The central task of Floquet theory is to calculate the quasienergy spectrum and the Floquet eigenstates of $\mathcal{H}(x, t)$. For this purpose, it is convenient to introduce a composite Hilbert space $\mathcal{R} \otimes \mathcal{T}$, where $\mathcal{R}$ is the Hilbert space of time-independent functions and $\mathcal{T}$ is the Hilbert space of functions with time periodicity T. We can choose the time-*independent* states $|\phi_m(x)\rangle$ satisfying $\langle\phi_n(x)|\phi_m(x)\rangle = \delta_{nm}$ as the orthonormal basis of space $\mathcal{R}$ and the time-*dependent* Fourier vectors $|e^{iq\Omega t}\rangle$ with $q = 0, \pm 1, \pm 2, \ldots$, as the orthonormal basis of space $\mathcal{T}$. Then, we can obtain the matrix of $\mathcal{H}(x, t)$ in the complete set of $\{|\phi_m(x)\rangle \otimes |e^{iq\Omega t}\rangle\}$ and calculate its quasienergy spectrum and the Floquet eigenstates.

Floquet Hamiltonian

From equation (2.124), the dynamics of Floquet eigenstate $|\psi_\alpha(x, t)\rangle$ at stroboscopic time moment $t = nT$ is given by

$$|\psi_\alpha(x, nT)\rangle = (e^{-i\epsilon_\alpha T})^n|\varphi_\alpha(x, 0)\rangle. \tag{2.126}$$

From the Schrödinger equation, the time evolution of the Floquet eigenstate $|\psi_\alpha(x, t)\rangle$ can be directly written in the Dyson form

$$|\psi_\alpha(x, nT)\rangle = \left(\mathcal{T}e^{-i\int_0^T H(t)dt}\right)^n \left|\psi_\alpha(x, 0)\right\rangle, \tag{2.127}$$

where $\mathcal{T}$ is the time-ordering operator. Comparing equation (2.126) to equation (2.127), the stroboscopic dynamics of the Floquet system can be described by an effective *time-independent* Hamiltonian H_F determined from

$$e^{-iH_F T} = \mathcal{T}e^{-i\int_0^T H(t)dt} \equiv U_f \tag{2.128}$$

or written formally as

$$H_F = i\frac{1}{T}\ln\left(\mathcal{T}\exp\left[-i\int_0^T H(t)dt\right]\right) \tag{2.129}$$

which has the eigenvalues ϵ_α and eigenstates $|\varphi_\alpha(x, 0)\rangle$. The time-independent Hamiltonian H_F is called *Floquet Hamiltonian.*

2.3.2 Magnus expansion

It is in general impossible to obtain a closed form for the Floquet Hamiltonian (2.129), except in very few examples like the driven harmonic oscillator. In the regime where the driving frequency is much larger than the characteristic frequency of the system, the Floquet Hamiltonian can be given in series functions, i.e., the *Floquet–Magnus expansion.*

By defining the time-evolution operator via $|\psi(x, t)\rangle = U(t, t_0)|\psi(x, t_0)\rangle$, we obtain the time-differential equation of motion for $U(t, t_0)$ from the Schrödinger equation, i.e.,

$$i\frac{\partial}{\partial t}U(t, t_0) = H(t)U(t, t_0).$$

The Magnus theorem [30, 31] claims that the time-evolution operator can be written in the form of $U(t, t_0) = \exp[-i\Omega(t, t_0)]$ with the operator $\Omega(t, t_0)$ determined by the following differential equation

$$\frac{d}{dt}\Omega(t, t_0) = \sum_{n=0}^{\infty}\frac{B_n}{n!}[\Omega^{(n)}, H], \tag{2.130}$$

where $[\Omega^{(n)}, H]$ is the multiple commutator defined via the recursion relationship $[\Omega^{(n)}, H] = [\Omega, [\Omega^{(n-1)}, H]]$ with $[\Omega^{(0)}, H] = H$, and B_n are Bernoulli numbers given by the identity $\frac{x}{e^x - 1} = \sum_{n=0}^{\infty}\frac{B_n}{n!}x^n$. The first few nonzero Bernoulli numbers are $B_0 = 1$, $B_1 = -1/2$, $B_2 = 1/6$, $B_4 = -1/30$ and $B_{2m+1} = 0$ for $m \geqslant 1$. We now replace H by εH in equation (2.130) and try a solution in the form of Magnus series

$$\Omega(t, t_0) = \sum_{n=1}^{\infty}\varepsilon^n\Omega_n(t, t_0). \tag{2.131}$$

By substituting the above series (2.131) into equation (2.130) and equating powers of ε, we can obtain in principle all the expansion orders $\Omega_n(t)$. For example, the explicit forms of the first three orders are given by [31]

$$\begin{aligned}
\Omega_1(t, t_0) &= \int_{t_0}^{t} dt_1 H(t_1),\\
\Omega_2(t, t_0) &= \frac{1}{2}\int_{t_0}^{t} dt_1\int_{t_0}^{t_1} dt_2[H(t_1), H(t_2)],\\
\Omega_3(t, t_0) &= \frac{1}{6}\int_{t_0}^{t} dt_1\int_{t_0}^{t_1} dt_2\int_{t_0}^{t_2} dt_3([H(t_1), [H(t_2), H(t_3)]]\\
&\quad + [[H(t_1), H(t_2)], H(t_3)]).
\end{aligned} \tag{2.132}$$

For the explicit expressions of higher orders, one can look in [32–36].

We apply the above Magnus theorem to the Floquet system. Because the Hamiltonian is a periodic function, we can decompose $H(t)$ in the form of Fourier harmonics

$$H(t) = \sum_{l \in \mathbb{Z}} H_l e^{il\Omega t} \quad \text{with} \quad \Omega = \frac{2\pi}{T}. \tag{2.133}$$

By setting the time-evolution interval $t \in [0, T]$, we have the Floquet Hamiltonian from equation (2.129) and the Magnus theorem

$$H_F = \frac{1}{T}\Omega(T, 0).$$

Letting the perturbative parameter $\varepsilon = 1$ in equation (2.131), we decompose the Floquet Hamiltonian in the form of Floquet–Magnus expansion $H_F = \sum_{n=0}^{\infty} H_F^{(n)}$. The first three orders series components of the Floquet Hamiltonian are given by [37]

$$\begin{aligned} H_F^{(0)} &= H_0 = \frac{1}{T}\int_0^T dt H(t), \\ H_F^{(1)} &= \frac{1}{\Omega}\sum_{l=1}^{\infty}\frac{1}{l}[H_l, H_{-l}], \\ H_F^{(2)} &= \frac{1}{\Omega^2}\sum_{l \neq 0}\left(\frac{[H_{-l}, [H_0, H_l]]}{2l^2} + \sum_{l' \neq 0, l}\frac{[H_{-l'}, [H_{l'-l}, H_l]]}{3ll'}\right). \end{aligned} \tag{2.134}$$

The leading term $H_F^{(0)} = H_0$ is nothing but the time average of the original Hamiltonian over one time period T.

Divergence problem

We should emphasise that, although the Magnus expansion method has a long history in physics, the precise radius of convergence of Magnus series is still an open mathematical problem. A sufficient condition for the convergence of Magnus series (2.131) in the time interval $t \in [0, t_c]$ is given by [31] $\int_0^{t_c} \left\| H(t) \right\| dt < \pi$, where the norm $\|A\|$ is defined as the square root of the largest eigenvalue of $A^\dagger A$, and the operator $H(t)$ needs to be a bounded operator in Hilbert space. More precise criterion of convergence can be found in [38]. For the Floquet–Magnus expansion, the convergence condition becomes $\int_0^T \left\| H(t) \right\| dt < \pi$ with $T = 2\pi/\Omega$ the time period. However, this convergence condition is not guaranteed in physical systems. For instance, since the generic non-integrable infinite-size systems eventually heat up to infinite temperature for an arbitrary driving frequency, it was concluded that the radius of convergence for the Floquet–Magnus expansion vanishes in the thermodynamic limit and the quantum ergodicity results in the divergence of the expansion [39–44]. It is still unclear whether the divergence of the Floquet–Magnus expansion necessarily results in the heat up to infinite temperature due to the difficulty to estimate the radius of convergence for non-integrable many-body systems. It was recently conjectured that the divergence of the Floquet–Magnus expansion is a universal feature of periodically driven nonlinear systems with energy localisation [45]. Nevertheless, since the system takes an exponentially long time to reach the infinite-temperature featureless state if the driving frequency is much larger than the local energy scale, the prethermal dynamics can be well described by a truncation of the Floquet–Magnus expansion [44, 46–52].

References

[1] Sussman G J and Wisdom J 2000 *Structure and Interpretation of Classical Mechanics* (Cambridge, MA: MIT Press)

[2] Landau L D and Lifshitz E M 1982 *Mechanics* 3rd edn (Oxford: Butterworth-Heinemann)

[3] Zheng Z and Hu G 2016 *From Dynamics to Statistical Physics* (Beijing: Peking University Publishing House)

[4] Deriglazov A 2010 *Classical Mechanics* (Berlin: Springer)

[5] Reichl L E 1992 *The Transition to Chaos: Conservative Classical Systems and Quantum Manifestations* (Berlin: Springer)

[6] Arnold V I (ed) 1988 *Dynamical Systems III* (Berlin: Springer)

[7] Goriely A 1996 Integrability, partial integrability, and nonintegrability for systems of ordinary differential equations *J. Math. Phys.* **37** 1871–93

[8] Arnold V I, Kozlov V V and Neishtadt A I 2006 *Mathematical Aspects of Classical and Celestial Mechanics* (Berlin: Springer)

[9] Arnol'd V 1963 Small denominators and problems of stability of motion in classical and celestial mechanics *Russ. Math. Surv.* **18** 85–191

[10] Moser J 1962 On invariant curves of area-preserving mapping of an annulus *Matematika* **6** 51–68

[11] Moser J 1962 On invariant curves of area-preserving mappings of an annulus *Nachr. Akad. Wiss. Gött., II. Math.-Phys. Kl.* **1962** 1–20

[12] Birkhoff G 1913 Proof of Poincaré's geometric theorem *Trans. Am. Math. Soc.* **14** 14–22

[13] Birkhoff G D 1927 *Dynamical Systems* vol 9 (New York: American Mathematical Society) p 9

[14] Birkhoff G D 1931 pr proof of the ergodic theorem *Proc. Natl Acad. Sci.* 17 656–60

[15] Birkhoff G D 1932 Probability and physical systems *Bull. Am. Math. Soc.* **38** 361–79

[16] Kolmogorov A N 1954 On the conservation of conditionally periodic motions under small perturbation of the Hamiltonian function *Dokl. Akad. Nauk. SSSR* **98** 527

[17] Moser J 1962 On invariant curves of area-preserving mappings of an annulus *Nachr. Akad. Wiss. Göttingen, Math.-Phys. Kl. IIa* 1–20

[18] Arnol'd V I 1963 Proof of a theorem of A. N. Kolmogorov on the preservation of conditionally periodic motions under a small perturbation of the Hamiltonian *Russ. Math. Surv.* **18** 9–36

[19] Arnold V I 1989 *Mathematical Methods of Classical Mechanics* 2nd edn (New York: Springer)

[20] Zaslavsky G M 2005 *Hamiltonian Chaos and Fractional Dynamics* (Oxford: Oxford University Press)

[21] Arnol'd V I 1964 Instability of dynamical systems with many degrees of freedom *Dokl. Akad. Nauk SSSR* **156** 9–12

[22] Givental A B, Khesin B A, Marsden J E, Varchenko A N, Vassiliev V A, Viro O Y and Zakalyukin V M (ed) 2009 *Instability of Dynamical Systems with Several Degrees of Freedom* (Berlin: Springer) pp 423–7

[23] Lochak P 1999 *Arnold Diffusion; A Compendium of remarks and Questions* (Dordrecht: Springer) pp 168–83

[24] Afanasiev V V, Chernikov A A, Sagdeev R Z and Zaslavsky G M 1990 The width of the stochastic web and particle diffusion along the web *Phys. Lett.* **A144** 229–36

[25] Floquet G 1883 Sur les équations différentielles linéaires à coefficients périodiques *Annales scientifiques de l'École Normale Supérieure* **12** 47–88

[26] Ince E L 1956 Ordinary *Differential equations* (New York: Dover)

[27] Shirley J H 1965 Solution of the Schrödinger equation with a Hamiltonian periodic in time *Phys. Rev.* **138** B979–87

[28] Winkler S and Magnus W 1979 *Hill's equation* (New York: Dover)

[29] Grifoni M and Hänggi P 1998 *Driven quantum tunneling Phys. Rep.* **304** 229–354

[30] Magnus W 1954 On the exponential solution of differential equations for a linear operator *Commun. Pure Appl. Math.* **7** 649–73

[31] Blanes S, Casas F, Oteo J A and Ros J 2009 The magnus expansion and some of its applications *Phys. Rep.* **470** 151–238

[32] Bialynicki-Birula I, Mielnik B and Plebanski J 1969 Explicit solution of the continuous Baker–Campbell–Hausdorff problem and a new expression for the phase operator *Ann. Phys.* **51** 187–200

[33] Mielnik B and Plebański J 1970 Combinatorial approach to Baker–Campbell–Hausdorff exponents *Ann. Inst. Henri Poincare* **12** 215–54

[34] Saenz L and Suarez R 2002 A combinatorial approach to the generalized Baker-Campbell-Hausdorff-Dynkin formula *Syst. Control Lett.* **45** 357–70

[35] Strichartz R S 1987 The Campbell-Baker-Hausdorff-Dynkin formula and solutions of differential equations *J. Funct. Anal.* **72** 320–45

[36] Suarez R and Saenz L 2001 Lie algebras associated with the exponential solutions of nonautonomous linear differential equations *J. Math. Phys.* **42** 4582

[37] Bukov M, D'Alessio L and Polkovnikov A 2015 universal high-frequency behavior of periodically driven systems: from dynamical stabilization to Floquet engineering *Adv. Phys.* **64** 139–226

[38] Casas F 2007 Sufficient conditions for the convergence of the magnus expansion *J. Phys. A: Math. Theor.* **40** 15001–17

[39] D'Alessio L and Polkovnikov A 2013 Many-body energy localization transition in periodically driven systems *Ann. Phys., NY* **333** 19–33

[40] Lazarides A, Das A and Moessner R 2014 Equilibrium states of generic quantum systems subject to periodic driving *Phys. Rev.* E **90** 012110

[41] D'Alessio L and Rigol M 2014 Long-time behavior of isolated periodically driven interacting lattice systems *Phys. Rev.* X **4** 041048

[42] Ponte P, Chandran A, Papić Z and Abanin D A 2015 Periodically driven ergodic and many-body localized quantum systems *Ann. Phys., NY* **353** 196–204

[43] Moessner R and Sondhi S L 2017 Equilibration and order in quantum Floquet matter *Nat. Phys.* **13** 424–8

[44] Bukov M, Heyl M, Huse D A and Polkovnikov A 2016 Heating and many-body resonances in a periodically driven two-band system *Phys. Rev.* B **93** 155132

[45] Haga T 2019 Divergence of the Floquet-Magnus expansion in a periodically driven one-body system with energy localization *Phys. Rev.* E **100** 062138

[46] Goldman N and Dalibard J 2014 Periodically driven quantum systems: Effective Hamiltonians and engineered gauge fields *Phys. Rev.* X **4** 031027

[47] Mori T, Kuwahara T and Saito K 2016 Rigorous bound on energy absorption and generic relaxation in periodically driven quantum systems *Phys. Rev. Lett.* **116** 120401

[48] Kuwahara T, Mori T and Saito K 2016 Floquet–Magnus theory and generic transient dynamics in periodically driven many-body quantum systems *Ann. Phys., NY* **367** 96–124

[49] Abanin D, De Roeck W, Ho W W and Huveneers F 2017 A rigorous theory of many-body prethermalization for periodically driven and closed quantum systems *Commun. Math. Phys.* **354** 809–27

[50] Abanin D A, De Roeck W, Ho W W and Huveneers F 2017 Effective Hamiltonians, prethermalization, and slow energy absorption in periodically driven many-body systems *Phys. Rev.* B **95** 014112

[51] Ho W W, Protopopov I and Abanin D A 2018 Bounds on energy absorption and prethermalization in quantum systems with long-range interactions *Phys. Rev. Lett.* **120** 200601

[52] Machado F, Kahanamoku-Meyer G D, Else D V, Nayak C and Yao N Y 2019 Exponentially slow heating in short and long-range interacting Floquet systems *Phys. Rev. Res.* **1** 033202

IOP Publishing

Phase Space Crystals
Condensed matter in dynamical systems
Lingzhen Guo

Chapter 3

Phase space lattices

In this chapter, we discuss the single-particle quantum physics in phase space lattices. Usually, the Hamiltonian of a particle is the sum of kinetic energy and potential energy. The kinetic energy is either parabolic (massive particle) or linear (massless particle) as a function of momentum. Therefore, there is no periodicity in the direction of momentum for a static Hamiltonian. However, when the particle is subject to a periodic drive, the effective Hamiltonian in the phase space of rotating frame can have lattice symmetry in both directions. In section 3.1, we introduce how to create Hamiltonians with various lattice structures in phase space. Based on the phase space lattice symmetry, a quantum band theory is built in section 3.2. Different fundamentally from real space, phase space is a noncommutative space with an intrinsic magnetic field. In section 3.3, we discuss the resultant topological physics in phase space and the application to bosonic code for fault-tolerant quantum computing.

3.1 Lattice symmetry in phase space

3.1.1 One-dimensional lattices

The theory of solid state physics is built on the discrete translation symmetry (lattice symmetry) of crystals in real space. The purpose of this book is to extend the study of solid state physics and condensed matter physics into phase space. Therefore, our preliminary task is to create lattice structures in phase space. Usually, the Hamiltonian of a time-independent system is parabolic (massive system) or linear (massless system) with respect to its momentum coordinate. As already discussed in section 2.2.5, the effective Hamiltonian of a periodically driven (Floquet) system can support some lattice structures in phase space.

Let us first consider the motion of a classical 1D system described by a time-independent Hamiltonian $H_0(x, p)$. In this case, the energy is conserved, and thus the orbit of the system in phase space is just the isoenergetic contour of $H_0(x, p)$. As discussed in section 2.2.1, we can define the *action* variable by

$$I \equiv \frac{1}{2\pi} \oint p \, \mathrm{d}x$$

along the isoenergetic contour and rewrite the Hamiltonian as a function of action $H_0(I)$. The conjugate variable of action I, i.e., the *angle* θ, does not explicitly appear in $H_0(I)$. According to the canonical equation of motion, we have $\theta(t) = \Omega t + \theta(0)$ with $\Omega(I) = dH_0(I)/dI$ representing the frequency of the regular motion orbit.

We then introduce a perturbation in the form of $H_1 = \epsilon h(x) f(t)$, where the time periodic function $f(t)$ is real and reads

$$f(t + 2\pi/\omega_d) = f(t) = \sum_k f_k e^{ik\omega_d t}$$

with $f_k = f_{-k}^*$. The spatial function $h(x)$ can be expressed with action and angle variables by $h(x) = \sum_l h_l(I) e^{il\theta}$. The whole single-particle Hamiltonian is then given by

$$H_s(t) = H_0(I) + \epsilon \sum_{l,k} h_l(I) f_k e^{il\theta + ik\omega_d t}. \tag{3.1}$$

If we tune the driving frequency such that it resonates with some rational orbit of action I_0, i.e., $\omega_d = n\Omega(I_0)$ ($n \in \mathbb{N}$), the motion of the particle in the vicinity of the resonant orbit can be separated into a fast oscillating and a slow varying part described by a new angle variable $\Theta = \theta - \omega_d t/n$. Using the generating function of the second kind $G_2 = I(\theta - \omega_d t/n)$, we transform into the rotating frame with frequency ω_d/n and retain the time-independent Hamiltonian from equation (3.1) in the rotating wave approximation (RWA), i.e.,

$$H_{\mathrm{RWA}}(I, \Theta) = H_0(I) - \frac{\omega_d}{n} I + \epsilon V_{\mathrm{eff}}(I, \Theta). \tag{3.2}$$

Here, the potential $V_{\mathrm{eff}}(I, \Theta) \equiv \sum_k h_{kn}(I) f_{-k} e^{ikn\Theta}$ is a periodic potential as a function of the angle Θ in the rotating frame. This effective Hamiltonian essentially describes a particle moving on the Θ ring in the presence of a time-independent effective lattice potential $V_{\mathrm{eff}}(I, \Theta)$ that can be engineered by tuning the Fourier components f_k and h_l in the experiments. For example, a monochromatic driving with frequency ω_d generates an effective lattice potential $V_{\mathrm{eff}}(I, \Theta) \propto \cos(n\Theta)$. According to the Poincaré–Birkhoff theorem in section 2.2.2, the driving resonates with the classical orbit with frequency ω_d/n and generates n identical local wells uniformly distributed on the Θ ring. We call such a 1D n-fold symmetric phase space lattice the $\mathbb{Z}_n$ *lattice*. Below, we introduce three concrete examples to illustrate the $\mathbb{Z}_n$ lattice structures in phase space.

3.1.1.1 Example I: power-law driven nonlinear oscillator

The first example is a nonlinear oscillator driven by an external field coupling to the oscillator's coordinate in a polynomial form. The resultant Hamiltonian is described by [1]

$$H_s(t) = \frac{p^2}{2m} + \frac{1}{2} m\omega_0^2 x^2 + \frac{\nu}{2} x^4 + 2f \cos(\omega_d t) x^n, \tag{3.3}$$

where ω_0 is the natural frequency of the oscillator and ω_d is the driving frequency. The degree of polynomial coupling is characterised by the exponent n. For the case of $n = 1$, the model (3.3) is the linearly driven Duffing oscillator [2–5]. For the case of $n = 2$, the model becomes the parametrically driven oscillator [6–8]. Recently, there is a lot of interest in the case of $n = 3$, i.e., the triply driven oscillator [9–13]. The model (3.3) extends the exponent n to be any integer number. In the recent experiment with superconducting circuits [14], the model has been realised up to $n = 5$.

In the subresonance regime, where the driving frequency ω_d is close to $n > 1$ times ω_0, i.e., $\delta\omega \equiv \omega_0 - \omega_d/n \ll \omega_0$, the time-independent RWA Hamiltonian with action and angle variables is given by [1]

$$H_{\mathrm{RWA}}(I, \Theta) = \left(I - \frac{1}{2}\right)^2 + \mu(\sqrt{2I})^n \cos(n\Theta). \tag{3.4}$$

Here, we have chosen the Harmonic oscillator as $H_0(x, p)$ and scaled the Hamiltonian by energy unit $4m^2\omega_0^2\delta\omega^2/3\nu$. The scaled driving strength is $\mu = \frac{3\nu f}{2}(-m\omega_0\delta\omega/3\nu)^{(n-4)/2}$ with red detuning $\delta\omega < 0$. We define two alternative conjugate variables

$$X \equiv \sqrt{2I}\cos\Theta, \quad P \equiv \sqrt{2I}\sin\Theta \tag{3.5}$$

and plot H_{RWA} in phase space spanned by X and P in figure 3.1(a). It is clear that n identical local wells distributed uniformly on a circle with radius $\sqrt{2I_0}$ with $I_0 = 1/2$. The RWA Hamiltonian (3.4) displays an n-fold rotational symmetry not visible in the original Hamiltonian (3.3). The n local wells only exist in a certain range of the driving strength $0 < |\mu| < \mu_c$ with the critical driving strength given by $\mu_c = (1 - r_c^2)/(nr_c^{n-2})$ with $r_c^2 = (n - 2)/(n - 4)$ [1].

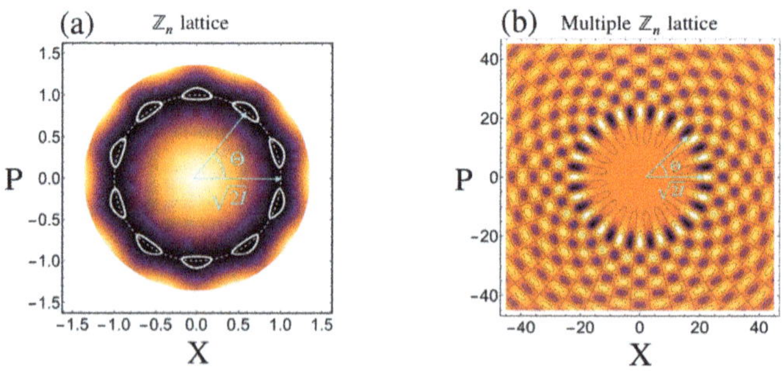

Figure 3.1. 1D lattice structures in phase space. (a) 2D density plot of Hamiltonian (3.4) in phase space for the power exponent $n = 10$ and driving strength $\mu = 0.3\mu_c$ with cut through the bottom of the quasienergy. The $\mathbb{Z}_n$ lattice has n stable states (white closed curves) arranged periodically in angular direction Θ, which is invariant under discrete phase space rotations $e^{i\Theta} \to e^{i(\Theta+\tau)}$ where $\tau = 2\pi/n$. (b) 2D plot of Hamiltonian (3.9) composed of many loops (white dashed circles) representing multiple $\mathbb{Z}_n$ lattices divided by the zero lines. There are $2n$ stable states arranged periodically in angular Θ direction but the lattice constant is still $\tau = 2\pi/n$, which means the whole lattice has two sublattices. Produced using own data but following methods in [3] and [19].

3.1.1.2 Example II: gravitational bouncer

The second example is a single atom with mass m bouncing in the vertical direction x on an oscillating hard-wall mirror with frequency ω_d, which is called *gravitational bouncer* [15–18]. There is a gravity field along the vertical direction with the gravitational acceleration g and a strong confinement in the transverse direction. The system is described by the following Hamiltonian

$$H_s = \frac{p^2}{2} + x[1 + \epsilon\cos(\omega_d t)] + V(x), \tag{3.6}$$

where the length, energy and time have been scaled dimensionless by the gravitational units $l_0 = [\hbar^2/(m^2 g)]^{1/3}$, $E_0 = mgl_0$ and $t_0 = \hbar/(mgl_0)$, respectively. Here, we have transformed from the laboratory frame to the oscillating frame of the mirror, where the effect of oscillating mirrors reappears as an inertial force modifying the gravitational acceleration periodically. The potential $V(x)$ describing the effect of hard-wall mirror is modelled by $V(x \leqslant 0) = +\infty$ and $V(x > 0) = 0$. By choosing $H_0 = p^2/2 + x + V(x)$ and using $H_0(I) = (3\pi I)^{2/3}/2$, the time-independent Hamiltonian (3.2) is given by [18]

$$H_{\mathrm{RWA}} = \frac{(3\pi I)^{2/3}}{2} - \frac{\omega_d}{n} I + \epsilon\left(\frac{3I}{\pi^2}\right)^{2/3}\frac{(-1)^{n+1}}{n^2}\cos(n\Theta). \tag{3.7}$$

The n: 1 resonant condition is fulfilled when the driving frequency is an integer multiple of the frequency $\Omega(I) = \partial H_0(I)/\partial I = [\pi^2/(3I)]^{1/3}$ of an unperturbed motion, i.e., $\omega_d = n\Omega(I_n)$, which implies $I_n = \pi^2 n^3/(3\omega_d^3)$. By introducing the conjugate variables (X, P) in equation (3.5), the plot of Hamiltonian (3.7) in the X–P plane is similar to figure 3.1(a).

3.1.1.3 Example III: cosine-driven harmonic oscillator

The third example is a harmonic oscillator driven by a cosine-type lattice potential with Hamiltonian described by [19]

$$H_s(t) = \frac{p^2}{2m} + \frac{1}{2}m\omega_0^2 x^2 + 2A\cos(kx + \omega_d t). \tag{3.8}$$

Under the subresonant condition $\omega_d/n = \omega$ with $n \in \mathbb{N}$, the RWA Hamiltonian in the classical limit with action and angle variables is given by [19]

$$H_{\mathrm{RWA}}(I, \Theta) = 2\mu J_n(\sqrt{2I})\cos(n\Theta - \frac{n\pi}{2}), \tag{3.9}$$

where $J_n(\bullet)$ is the Bessel function of order n. The Hamiltonian has been scaled by energy unit $m\omega^2/k^2$ and $\mu \equiv Ak^2/m\omega^2$ is the dimensionless driving strength.

In figure 3.1(b), we plot the scaled RWA Hamiltonian in phase space. The whole lattice structure clearly shows an n-fold rotational symmetry, i.e., a multiple $\mathbb{Z}_n$ lattice structure in phase space. The zero lines (i.e., $H_{\mathrm{RWA}} = 0$) divide the whole phase space lattice into many small 'cells'. The centre of each cell is a stable point corresponding to either a local minimum or a local maximum, which is called the *attractor* in the presence of dissipation. The area inside the cell represents the basin of the attractor in the cell's centre.

3.1.2 Two-dimensional lattices

In the above two examples, the periodic structure is one dimensional in the sense that the lattice symmetry appears only along the angle Θ direction in phase space. Is it possible to create a two-dimensional lattice in the whole phase space? The answer is yes. In section 2.2.5, we introduced the model of kicked harmonic oscillators (KHO) [20–23] with the Hamiltonian described by

$$H_s(t) = \frac{1}{2}(x^2 + p^2) + K\tau \cos x \sum_{n=-\infty}^{\infty} \delta\left(\frac{t}{\tau} - n\right), \tag{3.10}$$

where K is the dimensionless kicking strength and τ is the dimensionless kicking period. In the resonant condition that the kicking period satisfies $\tau = 2\pi/q_0$ with q_0 an integer and the weak kicking strength limit $|K| \ll 1$, the single-particle dynamics is still dominated by the fast harmonic oscillation but with slowly modulated amplitude and phase

$$x = P \sin\left(\frac{2\pi t}{\tau}\right) + X \cos\left(\frac{2\pi t}{\tau}\right), \quad p = P \cos\left(\frac{2\pi t}{\tau}\right) - X \sin\left(\frac{2\pi t}{\tau}\right). \tag{3.11}$$

The RWA Hamiltonian describing the slow dynamics of quadratures (X, P) is given by [24]

$$H_{\mathrm{RWA}}(X, P) = \frac{K}{q_0} \sum_{j=1}^{q_0} \cos\left(X \cos\frac{2\pi j}{q_0} + P \sin\frac{2\pi j}{q_0}\right). \tag{3.12}$$

In figure 3.2, we plot $H_{\mathrm{RWA}}(X, P)$ in phase space for different values of kicking number q_0. We see that the $H_{\mathrm{RWA}}(X, P)$ has a square lattice structure for $q_0 = 4$,

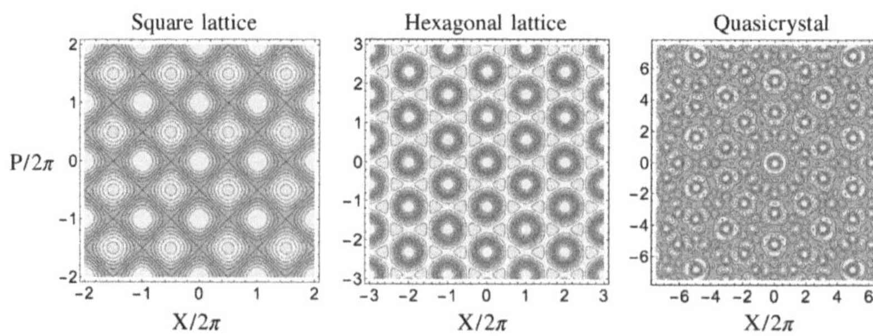

Figure 3.2. 2D phase space lattices from kicked harmonic oscillator: 2D density plot of Hamiltonian (3.12) showing (left) square lattice for $q_0 = 4$, (middle) hexagonal lattice for $q_0 = 3$ or $q_0 = 6$ and (right) quasicrystal structures for $q_0 = 5$. The value of $H_{\mathrm{RWA}}(X, P)$ has been scaled by the kicking strength K in all figures. Reprinted from [25].

hexagonal lattice structure for $q_0 = 3$ or $q_0 = 6$, and even quasicrystal structure for $q_0 = 5$ or $q_0 \geqslant 7$.

In order to synthesise more complex lattice structures in phase space, we consider the generalised model of kicked harmonic oscillator

$$H_s = \frac{1}{2}(x^2 + p^2) + \sum_{n\in\mathbb{Z}} \sum_{q} K_q \cos(k_q x - \phi_q)\delta\left(\frac{t}{\tau} - \theta_q - n\right), \tag{3.13}$$

where q represents the kicking sequence of stroboscopic lattices with tunable intensity K_q, wave vector k_q and phase ϕ_q at different time instances $t = \tau(n + \theta_q)$. For weak near-resonant driving ($|K_q| \ll 1, |\tau - 2\pi| \ll 1$), we obtain the general form of phase space Hamiltonian

$$H_s(X, P) = \frac{1}{2}\delta\omega(X^2 + P^2) + \sum_{q} K_q \cos\left[k_q(P\sin 2\pi\theta_q + X\cos 2\pi\theta_q) - \phi_q\right], \tag{3.14}$$

where we have defined the detuning $\delta\omega \equiv 1 - 2\pi/\tau$ between kicking and harmonic oscillation. We provide detailed derivation from equation (3.13) to equation (3.14) in appendix A. In principle, any arbitrary 2D lattice structures in phase space can be synthesised via multiple stroboscopic lattices.

3.1.2.1 Example: honeycomb lattice

As an example, we can generate a Hamiltonian with honeycomb lattice in phase space by setting twelve evenly spaced stroboscopic lattices, i.e., choosing $\theta_q = q/12$, with the kicking parameters: $K_q = \Lambda/2$, $k_q = 2$ and $\phi_q = 0$ for $q = 0, 2, 4, 6, 8, 10$; $K_q = \Lambda/2$, $k_q = 4/\sqrt{3}$ and $\phi_q = \pi$ for $q = 1, 5, 9$; $K_q = \Lambda$, $k_q = 2/\sqrt{3}$ and $\phi_q = \pi$ for $q = 3, 7, 11$. The resultant honeycomb lattice Hamiltonian [26] is (see appendix A for more details)

$$H_s(\boldsymbol{Z}) = -\frac{16}{9}\omega_0 \prod_{n=1}^{3} \sin^2\left(\frac{1}{2}\boldsymbol{v}_n \cdot \boldsymbol{Z}\right), \tag{3.15}$$

where we have defined the vector $\boldsymbol{Z} \equiv (X, P)$ and three ancillary vectors $\boldsymbol{v}_1 = (1, \frac{1}{\sqrt{3}})$, $\boldsymbol{v}_2 = (-1, \frac{1}{\sqrt{3}})$, $\boldsymbol{v}_3 = (0, -\frac{2}{\sqrt{3}})$. The extrema of $H_s(X, P)$ represent stable points of the classical dynamics. The small vibrations about these stable points have frequency $\omega_0 = -9\Lambda$ from the linear expansion of equation (3.15).

The two sublattices of the phase space honeycomb lattice Hamiltonian (3.15) have the same vibration frequency at their lattice sites. One can also tune the on-site frequencies of two sublattices with the following phase space lattice Hamiltonian

$$H_s^{\Delta} = -\frac{2\Delta}{\sqrt{3}} \sum_{n=1}^{3} \sin(\boldsymbol{v}_n \cdot \boldsymbol{Z}), \tag{3.16}$$

which can be implemented by additional stroboscopic lattices with kicking parameters $K_q = 2\Delta/\sqrt{3}$, $k_q = 2/\sqrt{3}$ and $\phi_q = 3\pi/2$ for $\theta_q = 1/12, 5/12, 9/12$. The on-site

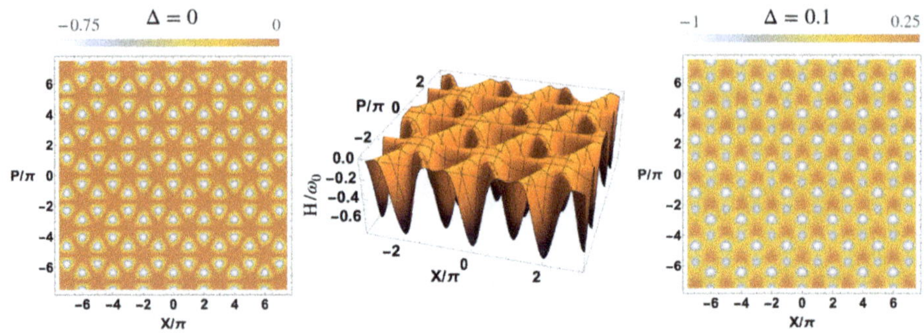

Figure 3.3. Honeycomb phase space lattice: 2D density plot (left) and 3D plot (middle) of the phase space honeycomb lattice Hamiltonian (3.15). The two honeycomb sublattices can be tuned asymmetric by setting the on-site detuning parameter Δ in equation (3.16) finite (right). Produced using own data but following the methods in [26].

frequencies of two sublattices are then $\omega_0 \pm 2\Delta$ with the *on-site detuning* parameter Δ. In figure 3.3, we plot the honeycomb lattice in phase space.

3.2 Quasinumber band theory

In this section, we calculate the energy spectrum of Hamiltonians with lattice symmetry in phase space. According to the Bloch theorem, a quasinumber band theory can be built for the $\mathbb{Z}_n$ phase space lattice. The parameters of tight-binding bands can be calculated by the method of Wentzel–Kramers–Brillouin (WKB) approximation.

3.2.1 Bloch theorem

Let us consider the example Hamiltonian (3.3). We transform into the rotating frame via unitary $\hat{U} = e^{i(\omega_d/n)\hat{a}^\dagger \hat{a} t}$, where $\hat{a}$ is the annihilation operator of the oscillator, resulting in $H_{\mathrm{RF}}(t) = \hat{U}H(t)\hat{U}^\dagger + i\hbar\dot{\hat{U}}\hat{U}^\dagger = H_{\mathrm{RWA}} + H_{\mathrm{nonRWA}}$. The RWA part Hamiltonian is given by (scaled by energy unit of $\epsilon_u = 4m^2\omega_0^2\delta\omega^2/3\nu$)

$$H_{\mathrm{RWA}} = -\lambda\hat{a}^\dagger\hat{a} + \lambda^2\hat{a}^\dagger\hat{a}(\hat{a}^\dagger\hat{a} + 1) + \frac{1}{2}\mu(2\lambda)^{\frac{n}{2}}(\hat{a}^{\dagger n} + \hat{a}^n). \tag{3.17}$$

Here, $\lambda \equiv -3\nu\hbar/(4m^2\omega_0^2\delta\omega)$ is the scaled dimensionless Planck constant, which describes the quantumness of our system, i.e., $[X, P] = i\lambda$. In the classical limit $\lambda \to 0$, the RWA Hamiltonian equation (3.17) goes back to the classical form equation (3.4).

The non-RWA part has the following explicit form [1]

$$\begin{aligned} H_{\mathrm{nonRWA}} = &\frac{1}{3}\lambda^2(2\hat{a}^\dagger\hat{a} - 1)\hat{a}^{\dagger 2}e^{i\frac{2\omega_d t}{n}} + \frac{1}{6}\lambda^2\hat{a}^{\dagger 4}e^{i\frac{4\omega_d t}{n}} \\ &+ \frac{1}{2}\mu(2\lambda)^{\frac{n}{2}}\left[\left(\hat{a}^\dagger e^{i\frac{2\omega_d t}{n}} + \hat{a}\right)^n - \hat{a}^n\right] + h.\ c. \end{aligned} \tag{3.18}$$

According to the Floquet–Magnus expansion (2.134), the RWA Hamiltonian H_{RWA} is just the leading order of Floquet Hamiltonian, i.e., $H_{\mathrm{RWA}} = H_F^{(0)} = T^{-1}\int_0^T H_{\mathrm{RF}}(t)\mathrm{d}t$ with Floquet period $T = n2\pi/\omega_d$. The non-RWA Hamiltonian contains many time-dependent terms with frequencies integer of $2\omega_d/n$. The RWA is valid as long as the motion from H_{RWA} is much slower than the motion from H_{nonRWA}, i.e., $|\epsilon_u/\hbar| < 2\omega_d/n$, which means the detuning has to be smaller than a critical point $|\delta\omega| < 2\lambda\omega_0$ [1].

For the cosine driving lattice model equation (3.8), the same transformation gives the following RWA Hamiltonian

$$H_{\mathrm{RWA}} = \mu\left[\left(\frac{\lambda}{2}\right)^{-\frac{n}{2}} e^{-\frac{\lambda}{4}-i\frac{n\pi}{2}}\hat{a}^n\, L_{\hat{a}^\dagger\hat{a}}^{-n}\left(\frac{\lambda}{2}\right) + h.\ c.\right]. \tag{3.19}$$

Here, functions $L_{\hat{a}^\dagger\hat{a}}^{-n}(\bullet)$ are the generalised Laguerre polynomials, as a function of the photon number $\hat{a}^\dagger\hat{a}|k\rangle = k|k\rangle$, where $|k\rangle$ are the Fock states. It can be proved that [19], in the classical limit $\lambda \to 0$, the RWA Hamiltonian equation (3.19) goes back to the classical form equation (3.9).

Both RWA Hamiltonians given by equation (3.17) and equation (3.19) display a new discrete symmetry not visible in their original Hamiltonians given by equation (3.3) and equation (3.8). In order to reveal this symmetry, we define a unitary operator $\hat{T}_\tau = e^{-i\tau\hat{a}^\dagger\hat{a}}$, which has the properties $\hat{T}_\tau^\dagger a\hat{T}_\tau = ae^{-i\tau}$ and $\hat{T}_\tau^\dagger a^n\hat{T}_\tau = a^n e^{-in\tau}$. It is direct to find that the RWA Hamiltonians in equations (3.17) and (3.19) are invariant under discrete transformation

$$T_\tau^\dagger H_{\mathrm{RWA}} T_\tau = H_{\mathrm{RWA}},\ \text{for}\ \ \tau = \frac{2\pi}{n}. \tag{3.20}$$

According to Bloch's theorem, the quasienergies $g(m)$ and eigenstates $\psi_m(\Theta)$ of the RWA Hamiltonian, satisfying $H_{\mathrm{RWA}}\psi_m(\Theta) = g(m)\psi_m(\Theta)$, have the form

$$\psi_m(\Theta) = \varphi_m(\Theta)e^{-im\Theta},\ \text{with}\ \ \varphi_m(\Theta + \tau) = \varphi_m(\Theta). \tag{3.21}$$

Here, we call the integer number m the *quasinumber* [1], since it is the quantised variable conjugate to the phase variable Θ similar to the role of the quasimomentum k in a crystal conjugate to the coordinate.

In figure 3.4, we plot the quasienergy band structures in the reduced Brillouin zone $m\tau \in [-\pi, \pi]$ from exact numerical diagonalisation of the $\mathbb{Z}_n$-lattice Hamiltonian (3.17) and the multiple $\mathbb{Z}_n$-lattice Hamiltonian (3.19). In the plot, we relabelled the eigenstates $\psi_m(\Theta)$ by $\psi_{lm}(\Theta)$, where $l = 1, 2, \ldots$ is the label of the bands counted from the bottom. The quasienergy band spectrum is discrete due to the finite values of n. It would become more continuous in the limit of large n.

3.2.2 Tight-binding model

For the quasienergy bands that are deep inside the lattice sites shown in figure 3.4, their band structures can be approximated by a tight-binding model

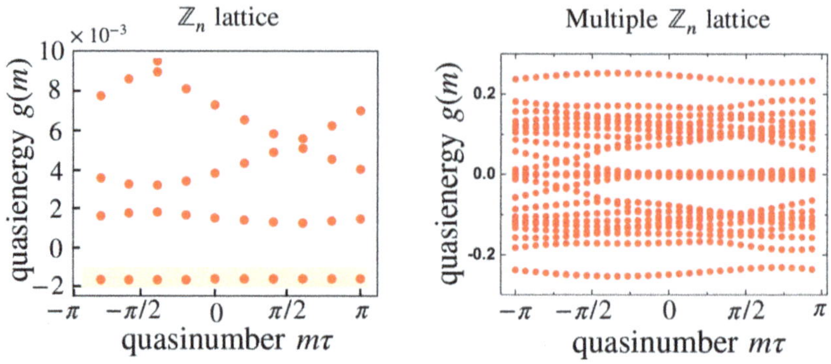

Figure 3.4. Quasienergy band structure in the reduced Brillouin zone $m\tau \in [-\pi, \pi]$. m is the quasinumber conjugate to the angle variable Θ. (Left) Z_n-lattice Hamiltonian (3.17) with parameters $\lambda = 1/205$, $n = 10$ and $\mu = 0.3\mu_c$; (Right) multiple Z_n-lattice Hamiltonian (3.19) with parameters $\lambda = 4$, $n = 30$ and $\mu = 1$. Produced using own data but following methods in [3] and [19].

$$H_{\mathrm{TB}} \approx E_l \sum_i \hat{a}_{i,l}^{\dagger}\hat{a}_{i,l} + \frac{1}{2}\left(J_l \sum_{\langle i,j\rangle} \hat{a}_{i,l}^{\dagger}\hat{a}_{j,l} + h.\,c.\right). \tag{3.22}$$

Here, $\hat{a}_{i,l}$ is the ladder operator of the lth Wannier at the ith lattice site $|\phi_{i,l}\rangle$. E_l is the energy level of the Wannier state, and J_l is the tunnelling rate of the Wannier state between nearest lattice sites denoted by $\langle i, j\rangle$. The tight-binding Bloch eigenstates and the localised Wannier states are related by the discrete rotation operator given in equation (3.20), i.e.,

$$|\psi_{lm}\rangle = \frac{1}{\sqrt{n}} \sum_{q=0}^{n-1} e^{imq\tau} T_\tau^q |\phi_{0,l}\rangle. \tag{3.23}$$

It can be checked that the above construction satisfies the required form of Bloch wave function given in equation (3.21), i.e., $T_\tau|\psi_{lm}\rangle = e^{im\tau}|\psi_{lm}\rangle$. The tunnelling rate is given by the overlap of two nearest-neighbour Wannier states

$$J_l = -\int_0^{2\pi} [T_\tau \phi_{0,l}(\Theta)]^* H_{\mathrm{RWA}} \phi_{0,l}(\Theta) \mathrm{d}\Theta, \tag{3.24}$$

where we have written the Wannier functions in the angle Θ representation. In general, the nearest-neighbour tunnelling rate is a complex number $J_l = |J_l|e^{i\delta_l\tau}$. From equations (3.23) and (3.24), we can obtain the spectrum of the lth band

$$\begin{aligned} g_l(m) &= \int_0^{2\pi} \psi_{lm}^*(\theta) H_{\mathrm{RWA}} \psi_{lm}(\theta) d\theta \approx E_l - J_l e^{im\tau} - J_l^* e^{-im\tau} \\ &= E_l - 2|J_l|\cos(m\tau + \delta_l\tau). \end{aligned} \tag{3.25}$$

The lth tight-binding band centres around E_l with bandwidth $d_l = 4|J_l|$. The parameter δ_l describes the asymmetry factor of lth band, which can be calculated by the WKB approximation.

3.2.3 Wentzel–Kramers–Brillouin method

There are three parameters in the tight-binding model to be determined, i.e., band central level E_l, tunnelling amplitude $|J_l|$ and tunnelling phase δ_l. The values of these parameters can be estimated by the Wentzel–Kramers–Brillouin (WKB) approximation which is valid in the semiclassical regime $\lambda \ll 1$.

Since the Z_n lattice has a circular discrete symmetry, it is convenient to consider the Hamiltonian H_{RWA} in the action–angle (I, Θ) coordinate system. Because I, Θ are conjugate variables and the canonical transformation (3.5) keeps the Poisson bracket invariant as discussed in section 2.1.3, we have the noncommutative relationship

$$[\Theta, I] = i\lambda. \tag{3.26}$$

The ladder operators of oscillator given by action–angle variables are $\hat{a} = e^{-i\Theta}\sqrt{I/\lambda}$ and $\hat{a}^{\dagger} = \sqrt{I/\lambda}\, e^{i\Theta}$. From the noncommutative relationship $[\hat{a}, \hat{a}^{\dagger}] = 1$, we have the identity $[I, e^{i\Theta}] = \lambda e^{i\Theta}$, which is consistent with equation (3.26). In the semiclassical limit $\lambda \ll 1$, the conjugate variables I and Θ define the phase space for our WKB calculation.

3.2.3.1 Band central level

In figure 3.5(left), we plot the $\mathbb{Z}_n$ lattice Hamiltonian equation (3.4) in the I–Θ plane. It displays n stable points surrounded by closed periodic orbits (solid white curves), and divided by separatrices (dashed white curves). In quantum mechanics, the classical orbit is quantised according to the so-called Bohr–Sommerfeld quantisation condition [27]

$$S(g) = \oint I d\Theta = 2\pi\lambda\left(k + \frac{1}{2}\right), \qquad k \in \mathbb{N}. \tag{3.27}$$

Here, the integral $S(g)$ represents the area enclosed by the classical orbit (isoenergetic contour of quasienergy g) and k takes non-negative integers. From the

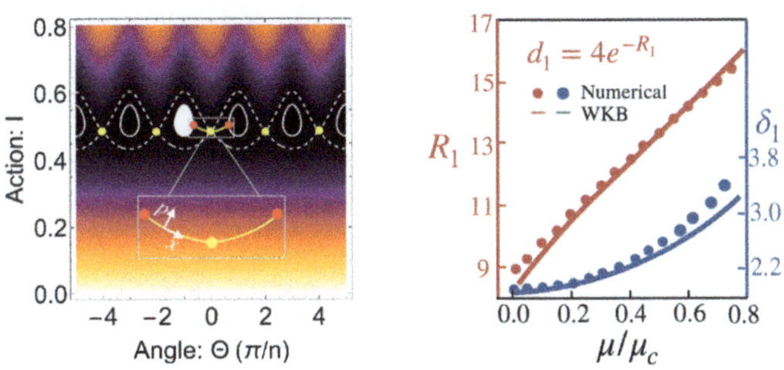

Figure 3.5. (Left) $\mathbb{Z}_n$-lattice Hamiltonian (3.4) in the action–angle I–Θ coordinate system with classical orbits (white closed curves), separatrices (white dashed curves), saddle points (yellow dots) and turning points (red points). The inset shows the local x–p coordinate along the dominant tunnelling path (yellow curve). (Right) Width of the lowest band d_1 and the asymmetry factor δ_1 versus driving for the band structure shown in figure 3.4(left). Produced using own data but following methods in [3].

above condition we can calculate the quasienergy levels $g = E_l$, which are the energy levels of Wannier states and the central levels of tight-binding bands.

3.2.3.2 Tunnelling amplitude

In order to calculate the tunnelling amplitude, we first need to determine the *turning points*. They are the edge points in the classically accessible regions, as indicated by the red points in figure 3.5(left). For a fixed quasienergy g, the action I is an implicit function of Θ given by the equation $H_{\mathrm{RWA}}(I, \Theta) = g$. The turning point of the angle Θ_a is the critical point where the action I becomes complex. The amplitude of J_l is given by the integral of the imaginary part of I in the classical forbidden region

$$|J_l| = \frac{\lambda}{2\pi}\left(\frac{dS}{dg}\right)^{-1}\Bigg|_{g=E_l} \exp\left(-\frac{1}{\lambda}\int_{\Theta_{a1}}^{\Theta_{a2}} |\mathrm{Im}[I]|\, d\Theta\right). \tag{3.28}$$

Here, $S(g)$ in the prefactor as a function of g is given by $S(g) = \oint I d\Theta$ in equation (3.27), and $\Theta_{a1(a2)}$ are the turning points of two neighbouring classical orbits.

In the semiclassical limit $\lambda \ll 1$, we can calculate approximately the tunnelling amplitude given by equation (3.28) along the dominant tunnelling path determined by the equation $\partial H_{\mathrm{RWA}}(I, \Theta)/\partial I = 0$. It can be proved that the dominant tunnelling path connects the turning points and the saddle points (see the definition in section 2.2.2) as indicated by the yellow line in figure 3.5(left). We set the curve along the dominant path as the 'coordinate' x axis and the direction perpendicular to the local curve segment as the 'momentum' p axis. Equivalently, we equip the dominant path with the local x–p coordinate system, which is an orthogonal rotation of the I–Θ coordinate basis as shown by the inset of figure 3.5(left). We write the Hamiltonian along the dominant path with the Taylor expansion to the second order in the local p direction

$$g(x, p) \approx \frac{1}{2}m_x\omega_x^2 p^2 + V(x), \tag{3.29}$$

where $m_x\omega_x^2 = \partial^2 H_{\mathrm{RWA}}/\partial p^2$. Note that the effective mass m_x and ω_x are not constant but depend on the position x of the dominant tunnelling path. The physics of Hamiltonian (3.29) can be interpreted as a particle in a double-well potential. Thus, the tunnelling amplitude can be calculated by the standard WKB theory [27]

$$|J_l| = \frac{\lambda\omega_e}{2\pi}\exp\left(-\frac{1}{\lambda}\int_{-x_a}^{x_a}\sqrt{\frac{2}{m_s\omega_s^2}[V(x) - E_l]}\,dx\right). \tag{3.30}$$

Here, we have set the centre of the x–p coordinate system on the saddle point between two turning points $\pm x_a$ that are given by $V(x_a) = E_l$. The property of saddle point is crucial for the tunnelling by modifying the exponent via $m_s\omega_s^2 = \partial^2 H_{\mathrm{RWA}}/\partial p^2|_{x=0}$, while the stable point only affects the prefactor via $\omega_e = (dS/dg)^{-1}|_{g=E_l}$.

In figure 3.5(right) we compare our approximate result for the first bandwidth $d_1 = 4|J_1|$ versus driving μ to numerical results for the $\mathbb{Z}_n$-lattice model (3.17) [1].

In the tight-binding regime they agree well with each other. A similar calculation for the band structure of the multiple $\mathbb{Z}_n$-lattice model (3.19) was provided in [19]. There, the situation is more complicated since the tunnelling between nearest $\mathbb{Z}_n$ lattices also plays an important role when bands cross each other.

The tunnelling rate $|J_l|$ decreases exponentially to zero in the classical limit of $\lambda \to 0$. This means the n eigenstates in each quasienergy band become degenerate and form a multiplet. In this limit, the Bloch eigenstate is unlikely to survive and will be broken into one of the localised Wannier states. In the lab frame, each broken state oscillates with time period of $2\pi n/\omega_d$, i.e., period multiplication in a quantum system. The $\mathbb{Z}_2$ symmetry breaking of parametrically driven model has been reported in the experiment with cold atoms [28] and was analysed in [29]. In the experiment with superconducting circuits [9, 14], the experimentalists observed robust output radiation (subharmonic oscillations [30]) at a frequency close to a fraction ω_d/n of the driving frequency and evenly $2\pi/n$-shifted phase components (up to $n = 5$). These observations call for further exploration of the quantum properties of the period multiplication and a search for engineering multicomponent quantum superpositions of macroscopic coherent states (multicomponent cat-states). While most of these experimental findings display classical fixed points and the hopping between them is induced by classical noise, the quantum mechanical coherences in the form of multicomponent cat-states are still absent.

3.2.3.3 Tunnelling phase

From the plot of the phase space lattices in figure 3.1, we would have expected a degeneracy $g(m) = g(-m)$, since clockwise and anticlockwise motion should be equivalent, as in the case of orbital motion. However, the corresponding band structures of phase space lattices shown in figure 3.4 exhibit a new feature that is not expected for the band structures of real space lattices. That is to say, the tight-binding bands show a surprising asymmetry described by the asymmetry factor δ_l.

As we have explained by equation (3.25), the asymmetry factor δ_l origins from the complex phase of the tunnelling rate $J_l = |J_l|e^{i\delta_l\tau}$ given by equation (3.24). In order to calculate the complex phase, we write the operator form of action I in the Θ representation from the noncommutative relationship (3.26)

$$I = \bar{I} - i\lambda\frac{\partial}{\partial\Theta}. \tag{3.31}$$

Here, we have introduced a constant $\bar{I}$ to guarantee the positivity of the action $\langle I\rangle > 0$. We then expand the phase translation operator $\hat{T}_\tau = e^{-i\tau\hat{a}^\dagger\hat{a}}$ by

$$\hat{a}^\dagger\hat{a} = \frac{1}{\lambda}I = \frac{1}{\lambda}\bar{I} - i\frac{\partial}{\partial\Theta}. \tag{3.32}$$

The operation of T_τ on the localised Wannier state $\phi_{0,l}(\Theta)$ is given by

$$\hat{T}_\tau\phi_{0,l}(\Theta) = e^{-i\bar{I}\tau/\lambda}\phi_{0,l}(\Theta - \tau). \tag{3.33}$$

According to equation (3.24), we obtain the symmetry factor

$$\delta_l = \frac{1}{\lambda}\bar{I}_l \ (\text{mod } n) \quad \text{with} \quad \bar{I}_l = \langle \phi_{0,l} | I | \phi_{0,l} \rangle. \tag{3.34}$$

The asymmetry factor δ_l is linearly dependent on the parameter $1/\lambda$ with the slope $\bar{I}_l$ (averaged action of the Wannier state) dependent on different tight-binding bands.

In figure 3.5(right), we show the dependence of the asymmetry factor δ_l on the driving strength μ for the $\mathbb{Z}_n$-lattice model, obtained both in the tight-binding calculation described above and from a numerical simulation. In the calculation, we have assumed $\bar{I}$ does not depend on the angle Θ, which means the dominant tunnelling path can be seen as a straight line. As a result, the symmetry factor δ_l is valid for large exponent $n \gg 1$. The curved feature of tunnelling path for small n can introduce significant additional correction, i.e., $\delta_l = \delta_l^0 + \bar{I}/\lambda(\text{mod } n)$. For example, more precise calculation for period tripling ($n = 3$) was done in [10].

3.2.4 Anderson localisation

In [31, 32], Sacha *et al* considered the problem of a single particle moving on a ring geometry under the action of a periodic force that, however, fluctuates randomly in time. The model is described by the following classical Hamiltonian,

$$H = (p - \alpha)^2/2 + V_0 g(\theta) f(t), \tag{3.35}$$

where the position of the particle on the ring is denoted by the angle variable θ and its canonical momentum by p. The Anderson problem is considered by introducing random disorder in space whereas here it is in time. Specifically, the time-dependent force consists of independent random Fourier components, e.g., $f(t) = \sum_{k=-\infty}^{\infty} f_k e^{ik\omega_d}$ with $f_k = f_{-k}^*$ independent random variables and ω_d the driving frequency. Note that the spatial potential $g(\theta) = \sum_k g_l e^{ik\theta}$ is not necessarily disordered but any regular function. Inside the classical invariant torus one can employ the secular approximation and eliminate the fast oscillating part of the motion and arrive at an effective Hamiltonian describing the slow motion,

$$H \approx \frac{P^2}{2} + V_0 \sum_{k \neq 0} g_k f_{-k} e^{ik\Theta} + \text{constant}. \tag{3.36}$$

Here P and Θ are canonical variables of the slow motion and they are related to the original ones by the transformation $\Theta = \theta - \omega_d t$ and $P = p - \alpha - \omega_d$. Assume the disordered potential satisfies the normal distribution $|g_k f_{-k}| \propto e^{-k^2/(2k_0^2)}$ with correlation length $\sqrt{2}/k_0$ and $\text{Arg}(f_{-k})$ are uniform random numbers in the interval $[0, 2\pi)$. With a suitable choice of V_0 and k_0, the eigenstates of (3.36) are Anderson localised even for energies higher than the standard deviation of the disordered potential. This is the Anderson localisation in phase space. This phenomenon can be revealed by measuring the probability that the particle passes a fixed position, which is localised exponentially around a certain moment of time. A possible experimental scheme for realising such Anderson localisation is an electron in a Rydberg atom perturbed by a fluctuating microwave field [33].

By introducing a random temporal perturbation satisfying $H'(t + nT_d) = H'(t)$ with $T_d = 2\pi/\omega_d$ to the driving field in (3.3) or (3.6) [34, 35], a random on-site potential in the tight-binding model can be produced, i.e.,

$$H_{\text{eff}} \approx -\frac{1}{2}\sum_{j=1}^{n}(J_j\hat{a}_{j+1}^{\dagger}\hat{a}_j + \text{c. c.}) + \sum_{j=1}^{n}\epsilon_j\hat{a}_j^{\dagger}\hat{a}_j. \tag{3.37}$$

Here, the random on-site energy

$$\epsilon_j = \int_0^{n2\pi/\omega_d} dt \int_0^{\infty} dz H'(t)|\phi_j|^2 \tag{3.38}$$

can be designed by properly choosing the fluctuating function in $H'(t)$. For the $\mathbb{Z}_n$ lattice, equation (3.37) is the 1D Anderson model on a ring. In the 1D Anderson model, the localisation always happens no matter how weak the disorder is. In the phase space picture, the localised state is the superposition of the wave-packets around the $\mathbb{Z}_n$ lattice site and with probability exponentially suppressed as the phase space distance increases. In the laboratory frame, the interpretation is that the probability density at a fixed position in the configuration space is localised exponentially around a certain moment of time [34], which is coined as the *Anderson localisation in time domain.*

3.2.5 Bloch oscillation

Bloch oscillation is a quantum phenomenon when a particle confined in a periodic lattice potential is subject to a weak constant force. It was first predicted by Felix Bloch while studying the motion of electrons in a solid crystal [36]. In this section, we will first discuss the Bloch oscillation in the real space, and then discuss how to realise such unusual phenomena in a $\mathbb{Z}_n$ phase-space lattice.

3.2.5.1 Infinite-size 1D lattice potential

Let us consider a particle, which has charge q and mass M, confined in an infinite 1D lattice potential $V(x) = V(x + a)$ with a the lattice constant. In the presence of a static electric field E along the coordinate direction, the total Hamiltonian is given by

$$H = \frac{\hat{p}^2}{2M} + V(x) - qEx \equiv H_0 - Fx. \tag{3.39}$$

Here, $F = qE$ is the constant force and $H_0 = \hat{p}^2/2M + V(x)$ is the lattice Hamiltonian. The eigenstates of H_0 are the Bloch states $\psi_{lk}(x) = e^{ikx}\varphi_{lk}(x)$ with $\varphi_{lk}(x) = \varphi_{lk}(x + a)$, where l is the band index and $k \in (-\frac{\pi}{a}, \frac{\pi}{a}]$ is the quasimomentum wavenumber.

For a given Bloch state $\psi_{lk_0}(x)$ as the initial state, we calculate the time evolution in the presence of constant force

$$\psi(t, x; E) = \exp\left(-\frac{i}{\hbar}\left[\frac{\hat{p}^2}{2M} + V(x) - qEx\right]t\right)\psi_{lk_0}(x). \tag{3.40}$$

It is not difficult to prove the time-evolution state $\psi(t, x; E)$ is still a Bloch state from the following relationship

$$\begin{aligned}\psi(t, x + a; E) &= \exp\left(\frac{1}{i\hbar}\left[\frac{\hat{p}^2}{2M} + V(x) - qEx\right]t\right)\exp\left(\frac{-qEa}{i\hbar}t\right) \\ &\quad \exp(ik_0 a)\psi_{lk_0}(x) \\ &= \exp\left[i\left(k_0 + \frac{qEt}{\hbar}\right)a\right]\psi(t, x; E).\end{aligned} \tag{3.41}$$

Here, in the first equality, we have used the lattice periodicity $V(x) = V(x + a)$ and the Bloch theorem $\psi_{lk_0}(x) = e^{ik_0 a}\psi_{lk_0}(x)$. From the second equality, we conclude that $\psi(t, x; E)$ is a Bloch state with quasimomentum wavenumber $k(t) \equiv k_0 + qEt/\hbar$. Equivalently, we have the following dynamics of quasimomentum wavenumber [37]

$$\frac{d}{dt}k(t) = \frac{q}{\hbar}E \quad \text{or} \quad \frac{d}{dt}(\hbar k) = F. \tag{3.42}$$

Strictly speaking, the Bloch state $\psi(t, x; E)$ is no more restricted to the initial band l as the constant force can mix Bloch states in different bands. That is to say, the Bloch state $\psi(t, x; E)$ is in general the superposition state $\psi(t, x; E) = \sum_{l'} c_{l'}\psi_{l'k(t)}(x)$. However, as long as the constant force is weak enough such that the inter-band transitions are suppressed by band gaps, we can restrict time evolution in the initial band, i.e., $\psi(t, x; E) \approx \psi_{lk(t)}(x)$.

One should distinguish the quasimomentum $\hbar k$ from the eigenvalue of momentum operator $\hat{p} = -i\hbar\partial/\partial x$. We define the semiclassical velocity of a Bloch state $\psi_{lk}(x)$ as [37]

$$v_l(k) \equiv \left\langle \psi_{lk}(x) \middle| \frac{\hat{p}}{M} \middle| \psi_{lk}(x) \right\rangle. \tag{3.43}$$

Since $\psi_{lk}(x)$ is assumed to be the eigenstate of H_0 with eigenenergy $E_l(k)$, we have

$$E_l(k) = \left\langle \psi_{lk}(x) \middle| \frac{\hat{p}^2}{2M} + V(x) \middle| \psi_{lk}(x) \right\rangle = \left\langle \varphi_{lk}(x) \middle| \frac{(\hat{p} + \hbar k)^2}{2M} + V(x) \middle| \varphi_{lk}(x) \right\rangle. \tag{3.44}$$

Here, we have used the property of Bloch state $\psi_{lk}(x) = e^{ikx}\varphi_{lk}(x)$ and the operator identity

$$e^{-ikx}\hat{p}e^{ikx} = \hat{p} + \hbar k. \tag{3.45}$$

Deriving with respect to the wavenumber k on both sides of the above equation, we have

$$\frac{dE_l(k)}{dk} = \left\langle \varphi_{lk}(x) \middle| \frac{\hbar(\hat{p} + \hbar k)}{M} \middle| \varphi_{lk}(x) \right\rangle. \tag{3.46}$$

In order to obtain the above identity, one should notice the two derivatives of wave functions result in $E_l(k)(d/dk)\langle\varphi_{lk}(x)|\varphi_{lk}(x)\rangle = 0$. According to equation (3.43), we have the semiclassical velocity [37]

$$v_l(k) = \frac{1}{\hbar}\frac{dE_l(k)}{dk}. \tag{3.47}$$

When considering a narrow Gaussian wave packet centred at k of l th band, the semiclassical $v_l(k)$ is just the group velocity of the wave packet.

According to equation (3.42), the period T_B that the particle completes the whole Brillouin zone is given by $2\pi/a = |qE|T_B/\hbar$, and the Bloch frequency is defined by

$$\omega_B = \frac{2\pi}{T_B} = \frac{|qE|a}{\hbar}. \tag{3.48}$$

For a tight-binding band (up to a constant) $E_l = J_l\cos(ka)$, the semiclassical velocity

$$v_l(t) = \frac{1}{\hbar}\left[\frac{dE_l(k)}{dk}\right]\Bigg|_{k=k_0+qEt/\hbar} = -\frac{J_l a}{\hbar}\sin\left[\left(k_0 + \frac{qEt}{\hbar}\right)a\right] \tag{3.49}$$

and the semiclassical position

$$x(t) = \int_{t_0}^{t} v_l(\tau)d\tau = x_0 + \frac{J_l}{qE}\cos\left[\left(k_0 + \frac{qEt}{\hbar}\right)a\right]. \tag{3.50}$$

As a result, the amplitude of Bloch oscillation in the position space is $A = |J_l/qE|$.

3.2.5.2 Finite-size loop lattice potential

Bloch oscillation can also be realised in a finite-size 1D loop lattice pierced by a time-dependent magnetic flux. We assume the loop lattice potential is $V(\Theta) = V(\Theta + \tau)$ with Θ the angular degree of freedom and $\tau = 2\pi/n$ the angular lattice constant. Supposing the the magnetic flux across the loop is $\Phi(t)$, the magnetic potential along the loop is $A(t) = \Phi(t)/2\pi r$ with r the radius of the loop. The Hamiltonian of the loop lattice is

$$H(t) = \frac{1}{2Mr^2}\left(\hat{p} - qAr\right)^2 + V(\Theta) \quad \text{with} \quad \hat{p} = -i\hbar\frac{\partial}{\partial\Theta} \tag{3.51}$$

We suppose the Bloch states in the absence of magnetic flux is $\psi_{lm}(\Theta) = e^{im\Theta}\varphi_{lm}(\Theta)$, i.e.,

$$H_0\left|\psi_{lm}(\Theta)\right\rangle = E_l(m)\left|\psi_{lm}(\Theta)\right\rangle \quad \text{with} \quad H_0 = \frac{1}{2Mr^2}\hat{p}^2 + V(\Theta). \tag{3.52}$$

Note that m is the quasimomentum wavenumber conjugate to the angle Θ. Using the operator identity (3.45), one can prove $\widetilde{\psi}_{lm(t)}(\Theta) \equiv e^{iqA(t)\Theta r/\hbar}\psi_{lm_0}(\Theta)$ is the temporal eigenstate of the time-dependent Hamiltonian $H(t)$, i.e.,

$$H(t)\left|\widetilde{\psi}_{lm(t)}(\Theta)\right\rangle = E_l(m_0)\left|\widetilde{\psi}_{lm(t)}(\Theta)\right\rangle. \tag{3.53}$$

Here we label the new temporal eigenstate with $m(t) = m_0 + qA(t)r/\hbar$ because $\widetilde{\psi}_{lm(t)}(x)$ is still a Bloch state with shifted quasimomentum wavenumber $m(t)$,

$$\widetilde{\psi}_{lm(t)}(\Theta + \tau) = e^{iqA(t)\Theta r/\hbar + iqA(t)r\tau/\hbar}e^{im_0\tau}\psi_{lm_0}(\Theta) = e^{im(t)a}\widetilde{\psi}_{lm(t)}(\Theta). \tag{3.54}$$

We now consider the time evolution of a Bloch state $\psi_{lm}(\Theta)$. If the time change of magnetic potential $A(t)$ is slow enough, one can prove the time evolution does not change the quasimomentum wavenumber m, i.e.

$$\begin{aligned}
\psi(t, \Theta; \delta A) &= \exp\left(-\frac{i}{\hbar}\left[\frac{1}{2Mr^2}(\hat{p} - q\delta Ar)^2 + V(\Theta)\right]t\right)\psi_{lm}(\Theta) \\
&= \exp\left(\frac{i}{\hbar}q\delta Ar\Theta\right)\exp\left(-\frac{i}{\hbar}\left[\frac{1}{2Mr^2}\hat{p}^2 + V(\Theta)\right]t\right) \\
&\quad \left[e^{-\frac{i}{\hbar}q\delta Ar\Theta}\psi_{lm}(\Theta)\right] \\
&\approx \exp\left[-\frac{i}{\hbar}E_l\left(m - \frac{q\delta Ar}{\hbar}\right)t\right]\psi_{lm}(\Theta)
\end{aligned} \tag{3.55}$$

Note that the state $e^{-\frac{i}{\hbar}q\delta Ar\Theta}\psi_{lm}(\Theta)$ is still a Bloch state with wavenumber $m - q\delta Ar/\hbar$, which is in general the superposition of Bloch states of H_0 in all the bands. To obtain the last step, we only retain the component from the l th band and neglect the inter-band transitions.

The time evolution adiabatically follows the temporal Bloch state

$$\widetilde{\psi}_{lm}(\Theta) = e^{iqA(t)r\Theta/\hbar}\psi_{l,m-qA(t)r/\hbar}(\Theta),$$

and the final state is

$$\psi(t, \Theta) = \exp\left[-\frac{i}{\hbar}\int_0^t E_l(m - qA(t)r/\hbar)dt + \frac{i}{\hbar}qA(t)\Theta r\right]\psi_{l,m-qA(t)r/\hbar}(\Theta). \tag{3.56}$$

Therefore, the time evolution under Hamiltonian $H(t)$ explores the Brillouin zone of H_0 via

$$m(t) = m - \frac{qA(t)r}{\hbar} \tag{3.57}$$

with accumulated dynamical phase and geometric phase. Comparing equation (3.57) to equation (3.42) and noticing $k = m/r$ for a loop geometry, we have $E = -dA/dt$ which is consistent with the Maxwell–Faraday law $\nabla \times \mathbf{E} = -\partial\mathbf{B}/\partial t$ and $\mathbf{B} = \nabla \times \mathbf{A}$ for a loop geometry. Following the proof of equation (3.47), the mechanical semiclassical angular velocity of time-evolution state $\psi(t, \Theta)$ is

$$\omega(t) \equiv \left\langle \psi(t, \Theta) \left| \frac{\hat{p} - qA(t)r}{Mr^2} \right| \psi(t, \Theta) \right\rangle = \frac{1}{\hbar} \frac{dE_l(m)}{dm}\bigg|_{m=m(t)}. \tag{3.58}$$

The mechanical semiclassical velocity along the loop is given by $v(t) = r\omega(t)$.

3.2.5.3 Bloch oscillation in $\mathbb{Z}_n$ phase-space lattice

We now discuss the Bloch oscillation in the $\mathbb{Z}_n$ phase-space lattice, cf. figure 3.1(a), which is described by the Hamiltonian (3.4). Using equation (3.31), the Hamiltonian for the $\mathbb{Z}_n$ phase-space lattice can be rewritten with action and angle variables as

$$H_n = \left(-i\lambda\frac{\partial}{\partial\Theta} + \bar{I} - \frac{1}{2}\right)^2 + \mu\left(\sqrt{2\bar{I}}\right)^n \cos(n\Theta). \tag{3.59}$$

Here, the average action value is determined by $\bar{I} = \bar{r}^2/2$ where $\bar{r}$ is the phase-space radius of the $\mathbb{Z}_n$ lattice. For the bottom band, the $\mathbb{Z}_n$ lattice radius is given by (see [1])

$$\begin{aligned} \bar{r} &= 1 - \lambda/2 + \sum\nolimits_{k=1}^{\infty} \bar{c}_{2k}\mu^{2k} \quad \text{with} \\ \bar{c}_{2k} &= \left(-\frac{n}{2}\right)^{2k} \frac{(2k-1)!![2k(n-2)-1]!!}{(2k)!!(2k)![2k(n-4)+1]!!}. \end{aligned} \tag{3.60}$$

Comparing Hamiltonian (3.59) to Hamiltonian (3.51), it is clear that the $\mathbb{Z}_n$ phase-space lattice is equivalent to a loop real-space lattice with effective magnetic flux (divided by 2π)

$$qAr = -\frac{1}{2}(\bar{r}^2 - 1) \approx \frac{1}{2}\lambda - \sum\nolimits_{k=1}^{\infty} \bar{c}_{2k}\mu^{2k}. \tag{3.61}$$

Here, we have used the semiclassical approximation $\lambda \ll 1$ and the weak driving $\mu \ll 1$. According to equation (3.61), one can tune the flux via the driving strength μ. From equation (3.57), the time evolution of an initial Bloch state $\psi_{lm}(\Theta)$ explores the Brillouin zone via

$$m(t) = m - \frac{1}{2} + \frac{1}{\lambda}\sum\nolimits_{k=1}^{\infty} \bar{c}_{2k}\mu^{2k} = m + \frac{1}{\lambda}(\bar{r} - 1). \tag{3.62}$$

If we tune the driving strength μ such that the radius $\bar{r}$ increases linearly, i.e., $\bar{r} = r_0 + \gamma t$, the Bloch frequency ω_B is determined by

$$\frac{\gamma}{\lambda}\frac{2\pi}{\omega_B} = n \quad \Longrightarrow \quad \omega_B = \frac{2\pi}{n\lambda}\gamma. \tag{3.63}$$

Using $\hat{a} = e^{-i\Theta}\sqrt{I/\lambda}$ and $\hat{a}^\dagger = \sqrt{I/\lambda}\, e^{i\Theta}$, cf. the discussion below equation (3.26), we can calculate the change of photon number for the time evolution state following the derivation of equation (3.58), i.e.,

$$\langle a^\dagger a\rangle(t) = \left\langle \psi(t,\Theta)\left|\frac{I}{\lambda}\right|\psi(t,\Theta)\right\rangle = \frac{1}{2\lambda} + \frac{1}{2\lambda^2}\frac{dE_l(m)}{dm}\bigg|_{m=m(t)}. \tag{3.64}$$

For a tight-binding band $E_l(m) = E_l - 2|J_l|\cos(m\tau + \delta_l\tau)$, cf. equation (3.25), we have

$$\langle a^\dagger a\rangle(t) = \frac{1}{2\lambda} + \frac{|J_l|\tau}{\lambda^2}\sin(m\tau + \delta_l\tau)\bigg|_{m=m(t)}. \tag{3.65}$$

Therefore, the maximum change of photon number is $\Delta\langle a^\dagger a\rangle_{max} = 2|J_l|\tau/\lambda^2$. Note that the bandwidth decreases exponentially as the driving strength increases, cf. figure 3.5(right). As a result, the Bloch oscillation of photon number, as a function of driving strength, is a sinusoid with an exponentially decaying envelope.

3.3 Noncommutative geometry

In the above section, we see that the tunnelling rate J_l of a $\mathbb{Z}_n$ lattice is a complex number in general. In physics, the complex tunnelling naturally arises in the plane of the phase space due to the noncommutative geometry. The phase space is fundamentally different from the real space as the two degrees of freedom of phase space do not commute, i.e., $[X, P] = i\lambda$ or $[\Theta, I] = i\lambda$. In the noncommutative phase space, the concept of point is meaningless. Instead, we should define a coherent state $|\alpha\rangle$ which is the eigenstate of the lowering operator, i.e., $\hat{a}|\alpha\rangle = \alpha|\alpha\rangle$ with $\hat{a} \equiv (X + iP)/\sqrt{2\lambda}$. We define the displacement operator by

$$D(\xi) \equiv \exp\left(\frac{\xi}{\sqrt{2\lambda}}\hat{a}^\dagger - \frac{\xi^*}{\sqrt{2\lambda}}\hat{a}\right), \tag{3.66}$$

which displaces the coherent state with an additional phase, i.e.,

$$D(\xi)|\alpha\rangle = e^{i\,\mathrm{Im}(\xi\alpha^*)}|\alpha + \xi\rangle. \tag{3.67}$$

As shown in figure 3.6, a coherent state $|\alpha\rangle$ is moved along the sides of a triangle in phase space by three displacement operators

$$D[-(\xi_1 + \xi_2)]D(\xi_2)D(\xi_1)|\alpha\rangle = e^{i\frac{1}{\lambda}S}|\alpha\rangle \tag{3.68}$$

with two complex numbers ξ_1, ξ_2 determining the moving path. The final coherent state naturally acquires a quantum phase factor $e^{iS/\lambda}$, where $S = \frac{1}{2}\,\mathrm{Im}(\xi_2\xi_1^*)$ is the

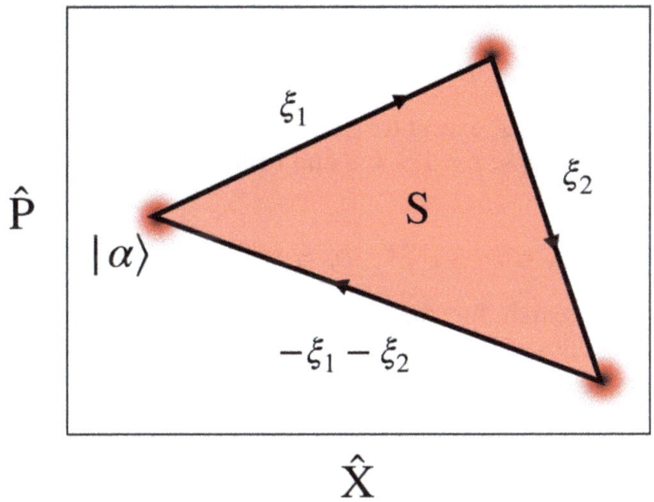

Figure 3.6. Geometric phase in noncommutative phase space. A coherent state $|\alpha\rangle$ is moved along a closed triangle and acquires a geometric phase factor given by $e^{i\frac{1}{\lambda}S}$ with S being the area of the enclosed path (red area). Reprinted from [24].

enclosed area [38]. This geometric phase is responsible for the asymmetry of quasienergy band structure, and also called Peierls phase [39] for describing the hopping of charged particles between tight-binding lattice sites in a magnetic field.

3.3.1 Hofstadter's butterfly

Topological phenomena in condensed matter physics have been intensively investigated in the past decades. In the quantum Hall physics, the origin of topology comes from the geometric phase induced by the applied magnetic field and the resultant band can be characterised by its topological invariant (Chern number or TKNN invariant) that relates to the quantised Hall conductance directly. In the noncommutative phase space, a geometric phase factor of a quantum state appears naturally when it moves along an enclosed path as shown in figure 3.6. We would expect that similar topological phenomena also exist in 2D phase space lattices.

As an example, we consider the Hamiltonian (3.12) in the case of $q_0 = 4$ which is a square lattice in form of

$$H_{\mathrm{sq}}(X, P) = \frac{1}{2}K(\cos X + \cos P). \tag{3.69}$$

This Hamiltonian is closely related to the established Harper's equation, which is a tight-binding model governing the motion of noninteracting electrons in the presence of a two-dimensional periodic potential and a uniformly threading magnetic field [39, 40]. The $H_{\mathrm{sq}}(X, P)$ is invariant under discrete translation operators in phase space defined by

$$\hat{D}_1 \equiv e^{i\frac{2\pi}{\lambda}P} \text{ and } \hat{D}_2 \equiv e^{i\frac{2\pi}{\lambda}X}. \tag{3.70}$$

However, the two translation operators $\hat{D}_1$ and $\hat{D}_2$ do NOT commute in general, i.e.,

$$[D_1^r, D_2^s] = D_1 D_2 (1 - e^{-i4\pi^2 rs/\lambda}) \tag{3.71}$$

with integer powers $r, s \in \mathbb{Z}$ except for $2\pi rs/\lambda \in \mathbb{Z}$. For $\lambda/2\pi = p/q$ with p and q two coprime integers, we choose $r = 1$, $s = p$ and define the following two commutative translational operators

$$D_X \equiv D_1 = e^{i\frac{2\pi}{\lambda}P}, \quad D_P \equiv D_2^p = e^{i\frac{2\pi p}{\lambda}X}. \tag{3.72}$$

Therefore, we can search for the common eigenstates of the complete set of commuting observables $\{D_X, D_P, H_{\text{sq}}\}$. We label the eigenvalues of operators D_X and D_P by $e^{i2\pi k_X}$ and $e^{i2\pi p k_P}$ respectively, and define the boundaries of the two dimensional Brillouin zone by $0 \leqslant k_X \leqslant 1$ and $0 \leqslant k_P \leqslant 1/p$, where k_X and k_P are called quasimomentum and quasicoordinate, respectively [41].

Due to the noncommutative geometry of phase space, i.e., $[X, P] = i\lambda$, we define two commutative translational operators along the X and P directions $T_X = e^{i\alpha P/\lambda}$ and $T_P = e^{i\beta X/\lambda}$ with the condition $\alpha\beta = 2\pi\lambda$. As illustrated in figure 3.6, the area enclosed by T_X and T_P with geometry phase $\alpha\beta/\lambda = 2\pi$ forms the primitive unit cell for phase space. Given the dimensionless Planck constant $\lambda/2\pi = p/q$, where p and q are coprime integers, we choose $\alpha = 2\pi/q$ and $\beta = 2\pi p$ for the translational operators. In the coordinate representation, we can choose the so-called Zak's basis [41] for the common eigenstates of T_X and T_P,

$$\phi_{(k_X,k_P)}(X) = \sqrt{\frac{\lambda}{q}} \sum_{l \in \mathbb{Z}} e^{ik_X \frac{2\pi l}{q}} \delta\left(X - \lambda k_P - \frac{2\pi l}{q}\right). \tag{3.73}$$

The Zak's basis (3.73) is composed of a series of Dirac's delta function with shifted phases. It is not difficult to check that $T_X|\phi_{(k_X,k_P)}(X)\rangle = e^{i2\pi k_X/q}|\phi_{(k_X,k_P)}(X)\rangle$ and $T_P|\phi_{(k_X,k_P)}(X)\rangle = e^{i2\pi p k_P}|\phi_{(k_X,k_P)}(X)\rangle$. Therefore, we can restrict the quantum numbers k_X and k_P in the region $0 \leqslant k_X < q$, $0 \leqslant k_P < 1/p$. Wave function (3.73) is also the eigenstate of the translation operator $D_X = T_X^q$ defined by equation (3.72) with the same quasimomentum k_X but different Brillouin range $0 \leqslant k_X < 1$. Therefore, eigenstates (3.73) with q different quasimomenta, i.e., $\phi_{(k_X,k_P)}(X)$, $\phi_{(k_X+1,k_P)}(X) = T_X \phi_{(k_X+1,k_P)}(X)$, $\cdots$ and $\phi_{(k_X+q-1,k_P)}(X) = (T_X)^{q-1}\phi_{(k_X+1,k_P)}(X)$, can be treated as q degenerate states of operator D_X, which guarantee the completeness of the Zak's kq representation introduced in [41].

In the presence of the square phase space lattice Hamiltonian $H_{\text{sq}}(X, P)$ given by equation (3.69), the q degenerate Zak's states for each (k_X, k_P) couple each other and the resultant quasienergy state of H_{sq} can be spanned as $|\psi_{b,\mathbf{k}}\rangle = \sum_{m=0}^{q-1} u_m |\phi_m(X)\rangle$, where $m \equiv (k_X + m, k_P)$ with $0 \leqslant k_X < 1$ and $\phi_m(X)$ is the basis (3.73) of the kq representation. In this subspace, the eigenequation is simply [24]

$$2\cos\lambda(k_X + m)u_m + e^{-i\lambda k_P}u_{m-1} + e^{i\lambda k_P}u_{m+1} = \frac{4E}{K}u_m \tag{3.74}$$

together with the boundary condition $u_m = u_{m+q}$. Therefore, given the parameters of $\lambda/2\pi = p/q$, k_X and k_P, the eigenvalues E of H_{sq} are determined by the following polynomial equation [24]

$$\cos(q\lambda k_X) + \cos(q\lambda k_P) = 1 + \frac{1}{2}\,\text{Tr}\prod_{j=1}^{q}\begin{bmatrix} \frac{4E}{K} - 2\cos(j\lambda) & -1 \\ 1 & 0 \end{bmatrix}. \tag{3.75}$$

The above eigenequation is a polynomial function of E with the degree of q. As a result, the whole band structure of H_{sq} is split into q sub-bands. Due to $[H_{\text{sq}}, D_2] = 0$, the state function $D_2^{\,n}|\psi_{b,\mathbf{k}}\rangle$ $(n = 0, 1, \ldots, p-1)$ is also the eigenstate of H_{sq}, whose quasimomentum and quasicoordinate are given by $(k_X + 2\pi n/\lambda)$ mod 1 and k_P respectively, and degenerate with eigenstate $|\psi_{b,\mathbf{k}}\rangle$. Therefore, each sub-band has p-fold degeneracy corresponding to p sequential 2π-shifted states along P direction. In summary, for each rational $\lambda/2\pi = p/q$ (remembering p, q are coprime integers), the spectrum contains q bands and each band has a p-fold degeneracy [42].

In figure 3.7(a), we plot the quasienergy spectrum for $\lambda/2\pi \in [0, 1]$, which shows a structure of Hofstadter's butterfly identical to that in quantum Hall systems. In figures 3.7(b), (d) and (e), we plot the quasienergy band structures in the two-dimensional Brillouin zone (k_X, k_p) for $\lambda/2\pi = 1/2$, $1/3$ and $2/3$, respectively. The quasienergy band structure is two-fold degenerate for $\lambda/2\pi = 2/3$ while there is no degeneracy for $\lambda/2\pi = 1/2$ and $\lambda/2\pi = 1/3$. In the case of $\lambda/2\pi = 1/2$ ($p = 1$ and $q = 2$), according to the general discuss above, the whole spectrum of H_{sq} has two non-degenerate sub-bands. For each $\boldsymbol{k} = (k_X, k_P)$, the eigenstate is $|\psi_{b,\mathbf{k}}\rangle = u_0|\phi_{(k_X,k_P)}(X)\rangle + u_1|\phi_{(k_X+1,k_P)}(X)\rangle$ with the eigenequation given by equation (3.74), i.e.,

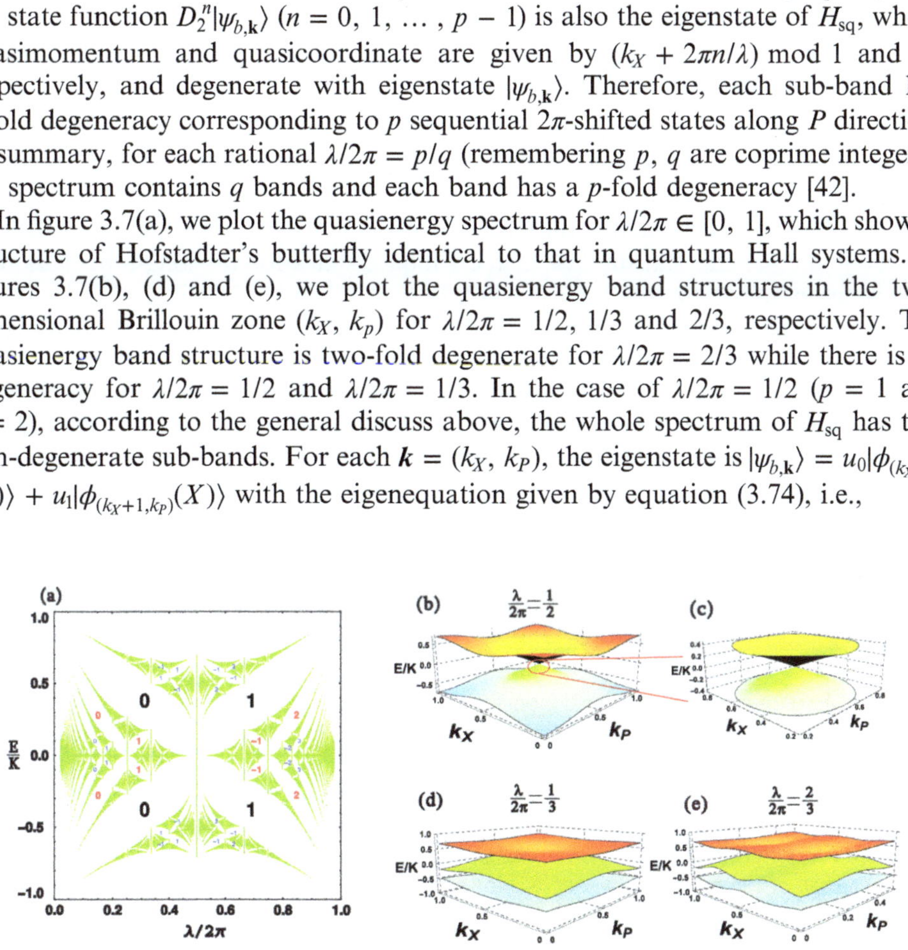

Figure 3.7. Topological band structures. (a) Hofstadter's butterfly: the quasienergy spectrum of Hamiltonian (3.69) with rational number $\lambda/2\pi \in [0, 1]$. The integers indicate the Chern number of each gap defined as the sum of Chern numbers of all the quasienergy bands below (or above) the gap. (b)–(e) Quasienergy band structures in 2D Brillouin zone for different parameters λ which are given above the figures. For $\lambda/2\pi = 1/2$, the two bands touch each other in the centre of the Brillouin zone. The linear dispersion relationship near the touching point is shown in figure (c). Reprinted from [24].

$$\begin{cases} u_0 \cos(\pi k_X) + u_1 \cos(\pi k_P) = \dfrac{2E}{K} u_0 \\ u_0 \cos(\pi k_P) - u_1 \cos(\pi k_X) = \dfrac{2E}{K} u_1. \end{cases} \tag{3.76}$$

The quasienergy band is given by $E = \pm\frac{K}{2}\sqrt{\cos^2(\pi k_X) + \cos^2(\pi k_P)}$, where the two sub-bands touch at the point $\boldsymbol{k} = (\frac{\pi}{2}, \frac{\pi}{2})$, see figure 3.7(b).

To show the underlying topology of the quasienergy bands, one can calculate the Chern number for a given band [43]

$$C_b = \frac{1}{2\pi} \oint_C \langle \psi_{b,\boldsymbol{k}} | \nabla_{\boldsymbol{k}} | \psi_{b,\boldsymbol{k}} \rangle \cdot \mathrm{d}\boldsymbol{k}, \tag{3.77}$$

where the contour C is integrated over the boundary of the Brillouin zone. The Chern number associated with a gap is subtle here since we are dealing with a Floquet system far from equilibrium. For the equilibrium systems, the Chern number of a gap is defined by the sum of the Chern numbers of the energy bands below the gap. However, the general statistic law of the Floquet states for the long-time stationary state is an on-going research topic [44–46]. In fact, the two sublattices of H_{sq} make no difference in the frame of Floquet theory. It is assumed that the statistic mechanics near the 'ground state' of each sublattice can be described by an effective Floquet–Gibbs statistics [46], i.e., $\rho = e^{-H_{\mathrm{sq}}/k_B T_{\mathrm{eff}}} / \operatorname{Tr}[e^{-H_{\mathrm{sq}}/k_B T_{\mathrm{eff}}}]$, where k_B is the Boltzmann constant and T_{eff} is the effective temperature in the rotating frame. It is shown that the effective temperature T_{eff} equals the real temperature if the local ground states near the stable points of phase space lattice are the standard coherent states [19]. Accordingly, the Chern number of a gap below (above) the zero energy line can be defined as the sum of Chern numbers of all the quasienergy bands below (above) the gap [24]. The Chern number of some gaps are calculated and labelled in figure 3.7(a).

Berry phase and Chern number Consider a quantum state $|\psi(\boldsymbol{k})\rangle$ specified by a set of parameters $\boldsymbol{k}$, e.g., $\boldsymbol{k} = \{k_X, k_P\}$ for the eigenstates of Hamiltonian (3.69). We define the *Berry connection* by

$$\boldsymbol{A} \equiv i\langle \psi(\boldsymbol{k}) | \nabla_{\boldsymbol{k}} | \psi(\boldsymbol{k}) \rangle, \tag{3.78}$$

where the quantum state is normalised $\langle \psi(\boldsymbol{k}) | \psi(\boldsymbol{k}) \rangle = 1$. The Berry connection describes the change of the relative phase $e^{-i\Delta\gamma}$ along a displacement segment in parameter space

$$\Delta\gamma = \boldsymbol{A} \cdot \Delta\boldsymbol{k} = -\operatorname{Arg}\left[\frac{\langle \psi(\boldsymbol{k}) | \psi(\boldsymbol{k} + \Delta\boldsymbol{k}) \rangle}{|\langle \psi(\boldsymbol{k}) | \psi(\boldsymbol{k} + \Delta\boldsymbol{k}) \rangle|}\right].$$

For a closed directed curve C in parameter space $\boldsymbol{k}$, the *Berry phase* along C is given by

$$\gamma(C) = \sum_i \Delta\gamma_i = \oint_C \boldsymbol{A} \cdot d\boldsymbol{k}. \tag{3.79}$$

Under the $U(1)$ gauge transformation $|\psi(\boldsymbol{k})\rangle \to e^{i\beta(\boldsymbol{k})}|\psi(\boldsymbol{k})\rangle$, the Berry connection is $\boldsymbol{A} \to \boldsymbol{A} + \nabla_{\boldsymbol{k}}\beta(\boldsymbol{k})$. Note that the Berry phase defined above is *gauge invariant* due to $\oint \nabla_{\boldsymbol{k}}\beta(\boldsymbol{k}) \cdot d\boldsymbol{k} = 0$. Chern number is defined by the Berry phase divided by 2π, as shown in equation (3.77), with the integral contour $\mathcal{C}$ over the boundary of the Brillouin. In the two dimensional parameter space $\boldsymbol{k} = (k_1, k_2)$, we consider a simply connected region $\mathcal{S}$ with oriented boundary $\partial\mathcal{S}$. Using the Stokes theorem, we have the Berry phase corresponding to $\partial\mathcal{S}$

$$\gamma(\partial\mathcal{S}) = \oint_{\partial\mathcal{S}} \boldsymbol{A} \cdot d\boldsymbol{k} = \int_{\mathcal{S}} (\partial_{k_1}A_{k_2} - \partial_{k_2}A_{k_1})dk_1dk_2 \equiv \int_{\mathcal{S}} B(k_1, k_2)dk_1dk_2, \tag{3.80}$$

where $B(k_1, k_2) \equiv \partial_{k_1}A_{k_2} - \partial_{k_2}A_{k_1}$ is called *Berry curvature*. In three-dimensional parameter space, the Berry curvature is given by $\boldsymbol{B}(\boldsymbol{k}) = \nabla_{\boldsymbol{k}} \times \boldsymbol{A}(\boldsymbol{k})$. A generation of Berry curvature in higher dimensional parameter space is given by [47, 48] $B_j = -\mathrm{Im}[\epsilon_{jkl}\partial_k A_l]$, where ϵ_{jkl} is the Levi–Civita symbol, and the repeating index represents Einstein summation. An efficient and gauge-invariant method of computing the Chern numbers in discrete Brillouin zone was provided in [49].

3.3.2 Chiral dissipative dynamics

The topology due to noncommutative phase space manifests itself also in the dynamics. In reality, the driven oscillators are inevitably in contact with the environment, which results in dissipation or decoherence of the quantum system. In the Born–Markovian approximation, the dissipative dynamics of the quantum KHO is described by the master equation [24],

$$\frac{d\rho}{dt} = -\frac{i}{\lambda}[H_s(t), \rho] + \kappa(n_0 + 1)\mathcal{D}[\hat{a}]\rho + \kappa n_0\mathcal{D}[\hat{a}^\dagger]\rho, \tag{3.81}$$

where κ characterises the dissipation rate and n_0 is the Bose–Einstein distribution of the thermal bath, and the Lindblad superoperator is defined by

$$\mathcal{D}[\hat{O}]\rho \equiv \hat{O}\rho\hat{O}^\dagger - \frac{1}{2}(\hat{O}^\dagger\hat{O}\rho + \rho\hat{O}^\dagger\hat{O}).$$

We emphasise that the above dissipative dynamics of KHO is based on the original full Hamiltonian (3.10) without RWA. To visualise the time evolution of quantum state, we define the Husimi Q-function of a given density matrix [50] $Q(\alpha, \alpha^*) \equiv \frac{1}{\pi}\langle\alpha|\rho|\alpha\rangle$, where $|\alpha\rangle$ is the coherent state. In [24], the time evolution of Husimi Q-function from the ground state of the harmonic oscillator was calculated, which is shown in figure 3.8. It is clearly seen that a final state with square lattice structure in phase space forms gradually revealing the underlying square structure of Hamiltonian (3.69).

Interestingly, the transient state shown in figure 3.8(b) has no reflection symmetries with respect to X and P although the RWA Hamiltonian (3.69) has. The quantum dynamics has a chiral feature as marked by the dashed lines along the backbone of the quasiprobability distribution. This chirality is a reflection of the

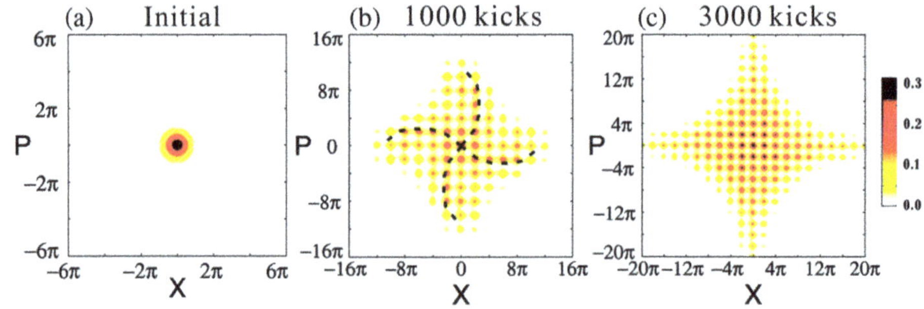

Figure 3.8. Chiral dissipative dynamics. (a) Husimi Q-function of the initial state, i.e., the ground state of the harmonic trapping potential. (b) Husimi Q-function after 1000 kicks with the main diffusion path showing a chiral feature (dashed lines). (c) Husimi Q-function after 3000 kicks. Parameters: $K = 0.1$ and $\kappa = 0.000\,1$. Reprinted from [24].

topological property of our system and the noncommutative geometry of the phase space. The probability amplitude of a particle appearing at a fixed point (a coherent state in phase space) is the sum of all the possible trajectories. Different from the path integral in 2D real space, each trajectory in phase space associates with a geometric phase due to the noncommutative geometry. The interference of geometric phase breaks the mirror symmetry of phase space. As approaching the stationary state, the chirality disappears in the end since the stationary state, which is a mixed state of the eigenstates of the RWA Hamiltonian (3.69), should recover the mirror symmetry of the RWA Hamiltonian. In the presence of dissipation, the diffusion process approaches a nonequilibrium stationary state with a finite size in phase space. The smaller the dissipation rate is, the lager the stationary phase space lattice is. To create a phase space lattice of large size, the dissipation rate κ has to be much weaker than the kicking strength $|K|$.

3.3.3 Gottesman–Kitaev–Preskill code

In this section, we discuss the eigenstates of $H_{\rm sq}(X, P)$ and their applications in fault-tolerant quantum computing. For $\lambda/2\pi = 1$, i.e., $p = 1$ and $q = 1$, the band structure is given by $E = \frac{K}{2}(\cos 2\pi k_X + \cos 2\pi k_P)$ from equation (3.74). For $K < 0$, the lowest-quasienergy eigenstate has quantum numbers $(k_X = 0, k_P = 0)$ and takes the form of the Zak's basis equation (3.73), i.e., $\phi_{(0,\,0)}(X) = \sum_{l\in\mathbb{Z}}\delta(X - 2\pi l)$. We relabel this eigenstate in an alternative form

$$|\bar{0}\rangle = \sum_{l\in\mathbb{Z}}|X = 2\pi l\rangle. \tag{3.82}$$

The comb-like state (3.82) is the coherent superposition of equally spaced eigenstates of position X, which was first introduced by D Gottesman, A Kitaev and J Preskill for bosonic quantum error correction [51]. In figure 3.9(a), we show this GKP state in the coordinate space and phase space. The other GKP state for qubit state encoding is given by

$$|\bar{1}\rangle = \sum_{l\in\mathbb{Z}} |X = \pi + 2\pi l\rangle, \tag{3.83}$$

which shifted by π of $|\bar{0}\rangle$ in the coordinate space. The two states $|\bar{0}\rangle$ and $|\bar{1}\rangle$ are the logical qubit states encoded in the infinite-dimensional Hilbert space of a system described by continuous quantum variables, i.e., the harmonic oscillator in our case.

3.3.3.1 Stabiliser operators

The GKP code exploits the noncommutative geometry of phase space and is resilient against the shift errors from the canonical variables X and P. To show the power of quantum error correction, we define two *stabiliser generators* that are used to detect the errors

$$S_X = e^{i2\pi P/\lambda}, \quad S_P = e^{i4\pi X/\lambda} \quad \text{with} \quad \lambda \equiv -i[X, P]. \tag{3.84}$$

The two stabiliser generators are commutative $[S_X, S_P] = 0$ for $\lambda = 2\pi$, and generate an Abelian stabiliser group, the *code stabiliser*. The *code space* is constructed by the simultaneous eigenspace of the commuting stabiliser generators. In our case, the code space is spanned by $|\bar{0}\rangle$ and $|\bar{1}\rangle$ which are eigenstates of both S_X and S_P with eigenvalue one. The code can correct errors in the set $\mathcal{E}$ if, for each pair of errors E_α and E_β, either $E_\alpha^\dagger E_\beta$ lies in the stabiliser group, or $E_\alpha^\dagger E_\beta$ fails to commute with some element of the stabiliser group [51]. If $E_\alpha^\dagger E_\beta$ commutes with all the code stabilisers but is not included in the stabiliser group, additional stabiliser generators are needed to reverse errors E_α and E_β. For example, when the logical qubit state is suffered by the shift error in the X direction $E_\alpha = e^{i\alpha P/\lambda}$, the corrupted state becomes $E_\alpha|\bar{n}\rangle = \sum_{l\in\mathbb{Z}}|X = \alpha + n\pi + 2\pi l\rangle$ $(n = 0, 1)$. The error can be diagnosed by measuring the stabiliser generator $S_P E_\alpha|\bar{n}\rangle = e^{i2\alpha}E_\alpha|\bar{n}\rangle$, which reveals the value of $2\alpha \bmod 2\pi$. For the same reason, the shift error in the P direction $E_\beta = e^{i\beta X/\lambda}$ can be diagnosed by measuring the stabiliser generator $S_X E_\beta|\bar{n}\rangle = e^{i\beta}E_\beta|\bar{n}\rangle$. The errors can be reversed by applying either $E_\alpha^\dagger$ and $E_\beta^\dagger$, which are actually the displacement operators defined by equation (3.66).

3.3.3.2 Clifford gates

The logical operators acting on the encoded qubit have to commute with the code stabiliser but are not contained in the stabiliser group. In our case, we define the logical operators as following

$$\sigma_X = e^{i\pi P/\lambda}, \quad \sigma_Z = e^{i2\pi X/\lambda}, \quad \sigma_Y = e^{i\pi P/\lambda + i2\pi X/\lambda} \quad \text{with} \quad \lambda = 2\pi. \tag{3.85}$$

One can check that logical operators σ_X, σ_Y and σ_Z satisfy the anti-commutative relationship $\{\sigma_\alpha, \sigma_\beta\} = 0$ for $\alpha \neq \beta$ with $\alpha, \beta \in \{X, Y, Z\}$ using $e^A e^B = e^{[A, B]}e^B e^A$ (which holds if A and B commute with their commutator). When acting on the logical qubit states, we have $\sigma_Z|\bar{n}\rangle = (-1)^n|\bar{n}\rangle$, $\sigma_X|\bar{n}\rangle = |1 - \bar{n}\rangle$ and $\sigma^2_{\alpha=X, Y, Z} = 1$. We illustrate the logical operators in figure 3.9(b), and the eigenstates of σ_X, i.e., $|\bar{0}\rangle \pm |\bar{1}\rangle$, in the lower panel of figure 3.9(a).

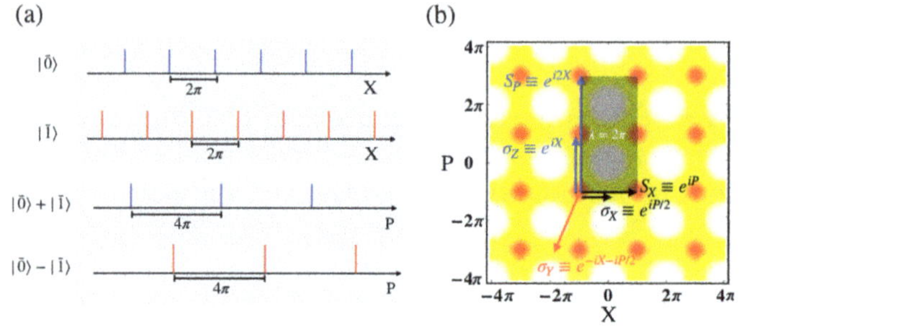

Figure 3.9. Gottesman–Kitaev–Preskill (GKP) code. (a) Two GKP code states $|\bar{0}\rangle$, $|\bar{1}\rangle$ and their Fourier-transformed states $|\bar{0}\rangle \pm |\bar{1}\rangle$. They are coherent superposition states of 2π-spaced position eigenstates (upper panel) or 4π-spaced momentum eigenstates (lower panel). (b) The colour plot is the Husimi Q-function of GKP state $|\bar{0}\rangle$ with two code stabilisers S_X, S_P to detect errors and three Pauli operators σ_X, σ_Y and σ_Z for logical operations.

Above, we have discussed the logical operations and error corrections in single-qubit encoded space. In order to perform universal quantum computing, we need to couple multiple logical qubits and implement universal quantum gates. In general, a set of complete universal quantum gates include ‘easy’ Clifford gates, which form the *Clifford group*, and ‘difficult’ gates that are not in the Clifford group. To explain the Clifford group, we first construct the so-called *Pauli group* for N-qubit system

$$\mathbb{P}_N = \left\{ e^{i\theta\pi/2}\sigma_{\alpha_1} \otimes \cdots \otimes \sigma_{\alpha_N} \middle| \theta = 0,\, 1,\, 2,\, 3;\, \alpha_i = I,\, X,\, Y,\, Z \right\}, \tag{3.86}$$

where σ_I represents the identity operator. The Clifford group $\mathbb{C}_N$ is defined as a set of 2^N-dimensional unitary transformations that preserve the ‘Pauli group’, i.e.,

$$\mathbb{C}_N = \left\{ V \in U(2^N) \,\middle|\, V\mathbb{P}_N V^\dagger = \mathbb{P}_N \right\}. \tag{3.87}$$

Since each σ_Y gate is a product of σ_X and σ_Z gates (up to a constant phase), it suffices to judge a unitary U lies in the Clifford group by $UXU^\dagger \in \mathbb{P}_N$ and $UZU^\dagger \in \mathbb{P}_N$. In our case, Pauli operators can be expressed in terms of canonical variables X_i and P_i in the form of

$$U_{\alpha\beta} = \exp\left[i \sum_{i=1}^{N} (\alpha_i P_i + \beta_i X_i) \right], \tag{3.88}$$

where the α_i and β_i are real numbers. The Clifford group transformations are the linear transformations of the X_i and p_i that preserve the canonical commutation relations, which are called *symplectic transformations*. The symplectic transformations and thus the gates in the Clifford group can be accomplished using linear optical elements (phase shifters and beam splitters), along with elements that can ‘squeeze’ an oscillator [51]. Any element in the Clifford group is generated by three basic gates: Hadamard gate (H), phase gate (S) and controlled-NOT gate ($CNOT$), which are defined as

- H: $H\sigma_X H^\dagger = \sigma_Z$, $H\sigma_Z H^\dagger = \sigma_X^{-1}$

- S: $S\sigma_X S^\dagger = \sigma_Y$, $S\sigma_Z S^\dagger = \sigma_Z$
- $CNOT$: $CNOT\sigma_{X_1}^{n_1}\sigma_{X_2}^{n_2}CNOT^\dagger = \sigma_{X_1}^{n_1}\sigma_{X_2}^{n_2-n_1}$, $CNOT\sigma_{Z_1}^{n_1}\sigma_{Z_2}^{n_2}CNOT^\dagger = \sigma_{Z_1}^{n_1+n_2}\sigma_{Z_2}^{n_2}$.

Here, the $CNOT$ gate is a two-qubit gate with qubit 1 the control of gate and qubit 2 its target. For the Pauli operators defined by equation (3.85), the H, S and $CNOT$ gates are

- H: $X \to -P/2$, $P \to 2X$
- S: $X \to X$, $P \to P + 2X$
- $CNOT$: $X_1 \to X_1$, $P_1 \to P_1 - P_2$; $X_2 \to X_2 + X_1$, $P_2 \to P_2$.

3.3.3.3 Universal quantum gates

For Clifford gate operation and error recovery, it is adequate to use symplectic transformations together with homodyne measurements. However, the Clifford gates do not form a universal set of quantum gates because some gates cannot be arbitrarily approximated with a finite set of symplectic Clifford operations. A typical non-symplectic gate is the $\pi/8$-*phase gate* given by $S_{\pi/8} = e^{-i\pi/8}\sqrt{S}$, where S is the symplectic phase gate defined above. Fortunately, the non-symplectic $S_{\pi/8}$ phase gate can be realised by photon counting and thus completes a universal set of fault-tolerant quantum gates [51].

3.3.3.4 Quantum error detection and recovery

In order to diagnose errors in a coded data state, we must measure the stabiliser generators S_X and S_P. But how do we implement such measurement in the experiment? A general procedure proposed by Steane [52] is to 'feed' the errors from the code block to a coded ancilla using $CNOT$ gate, and then perform a homodyne measurement of the X or P quadrature of the ancilla. For example, to measure the stabiliser operator $S_X = e^{i2X}$ (i.e., the value of X mod π), we prepare the ancilla in the state $(|\bar{0}\rangle + |\bar{1}\rangle)/\sqrt{2} = \sum_{l\in\mathbb{Z}}|X = \pi l\rangle$ and execute the $CNOT$ gate with the data as control and the ancilla as the target. According to the symplectic transformation of the $CNOT$ gate, we have $X_2 \to X_2 + X_1$, where X_1, X_2 are the quadratures for the data and ancilla before the execution of the $CNOT$ gate. The value of $X_1 + X_2$ can be obtained by destructively measuring the X of ancilla, and this value modulo π determines the shift that should be applied to the data for error recovery. A similar measurement of the stabiliser generator $S_P = e^{iP}$ can be performed and the shift errors of quadrature P can be diagnosed and corrected [51].

In [51], the authors discussed three possible error channels which can be protected against by GKP code. First, the GKP quantum error-correcting code can successfully suppress decoherence caused by quantum diffusion, which is described by the master equation

$$\dot{\rho} = -\frac{D}{2}[X, [X, \rho]] - \frac{D}{2}[P, [P, \rho]],$$

if the recovery operation is applied frequently enough. That is to say roughly, the time interval between error correction steps should be small compared to the characteristic diffusion time D^{-1}. Second, the GKP error correction is also effective to recover the amplitude damping of the oscillator, which is described by the master equation

$$\dot{\rho} = \Gamma(a\rho a^\dagger - \frac{1}{2}a^\dagger a\rho - \frac{1}{2}\rho a^\dagger a),$$

if the time interval between error correction steps is small compared to the damping time Γ^{-1}. Third, except the decoherence, the GKP code can also protect against *unitary errors*, e.g., an inadvertent phase rotation induced by the unitary transformation $U(\theta) \equiv e^{i\theta a^\dagger a}$. Because such a rotation by a small angle θ can be accurately expanded in terms of small displacements, the GKP error correction is effective if an oscillator is slightly overrotated or underrotated.

3.3.3.5 GKP states preparation

There are three important ingredients for fault-tolerant quantum computation: (1) prepare encoded states; (2) perform error recovery; (3) execute universal quantum gates. We have discussed how to execute universal quantum computation fault tolerantly, and how to perform quantum error correction. However, the most difficult task is to prepare encoded states. In the syndrome measurements of errors, we need ancilla prepared in the encoded space. In order to prevent small errors from accumulating and eventually damaging the encoded quantum information, a fresh supply of encoded ancilla oscillators needs to be continuously available.

Because the symplectic transformations preserve Gaussian (coherent or squeezed) states of continuous variables, the Clifford gates cannot take the oscillator's ground state to a GKP encoded state. Therefore, GKP encoding requires non-symplectic operations that resort to a nonlinear coupling between oscillators. In [51], the authors proposed to achieve such purpose by coupling the data oscillator to another metre oscillator via the following interaction Hamiltonian $H_I = \epsilon X(b^\dagger b)$, where $b^\dagger(b)$ is the ladder operator of the metre oscillator. As a consequence, the frequency of the metre oscillator is modified by $\Delta\omega_{\text{meter}} = \epsilon X$. If this coupling is turned on for a time $t = 1/\epsilon$, the phase of the metre oscillator is advanced by $\Delta\phi = X$. By preparing the metre oscillator in the eigenstate of the momentum $P = 0$ and measuring its advanced phase, one can determine the value of X mod 2π for the data oscillator, and apply a shift if necessary to obtain the state with $|\bar{0}\rangle$ or $|\bar{1}\rangle$. Furthermore, the many-oscillator encoded states can be achieved with symplectic transformations, i.e., linear optical operations and squeezing. In the recent experiments [53, 54], the metre oscillator was replaced by an ancillary physical qubit, i.e., an internal two hyperfine levels for trapped ion [53] or a superconducting qubit (transmon) for cavity mode [54]. The quadrature of the data oscillator is coupled to the Pauli operator of the ancillary physical qubit. The GKP encoded states are prepared by detecting the state of ancilla qubit, i.e., the modular variable measurements. However, the *active* implementation of GKP code requires complicated protocols. To diagnose the errors, we must continually measure the stabiliser operators to detect error syndrome. To recover code, we must frequently perform error correction by applying shifts of the quadratures conditioned on the measurement outcomes. High-quality ancilla oscillators must be provided since they are steadily consumed by the syndrome measurements.

An alternative method to avoid this complex scheme is the *passive* control, in which the protection is a built-in feature of the system's physical dynamics. For example, we can engineer a quantum system whose ground state is the GKP code subspace. The system will relax to the code space by the natural coupling to its dissipative environment. The associated energy penalty of errors protects the system automatically relaxing back into the computational space. In [51], the authors proposed to realise such physical fault tolerance by the codes for an electron in a Landau level together with a tunable periodic potential whose unit cell encloses two flux quanta (corresponding to $\lambda = 4\pi$ in our language). In this case, the Landau levels are split into two energy bands. By turning on the potential and waiting for the electrons to relax to the bottom of the lowest band, an encoded state could be prepared. In [55], the authors proposed a superconducting circuit design of Landau levels in a crystal potential, composed of two Josephson junctions coupled by the *gyrator* (a nonreciprocal element). They investigated the case of half flux quanta contained in the crystal unit cell (corresponding to $\lambda = \pi$ in our language), and the code space is spanned by the two-fold degenerate ground states. The GKP code in such a circuit is naturally protected against the common noise channels such as charge and flux noise. The phase space lattice method introduced here may provide another way to dynamically generate a passively stable GKP code without explicit active stabiliser measurements, which can be implemented by an LC oscillator comprising a *superinductance* in parallel to a Josephson junction [56].

References

[1] Guo L, Marthaler M and Schön G 2013 Phase space crystals: a new way to create a quasienergy band structure *Phys. Rev. Lett.* **111** 205303

[2] Dykman M I 1975 Theory of nonlinear nonequilibrium oscillators interacting with a medium *Zh. Eksp. Teor. Fiz.* **68** 2082–94

[3] Peano V and Thorwart M 2006 Dynamics of the quantum duffing oscillator in the driving induced bistable regime *Chem. Phys.* **322** 135–43 (Real-time dynamics in complex quantum systems)

[4] Guo L-Z, Zheng Z-G and Li X-Q 2010 Quantum dynamics of mesoscopic driven duffing oscillators *Europhys. Lett.* **90** 10011

[5] Guo L, Zheng Z, Li X-Q and Yan Y 2011 Dynamic quantum tunneling in mesoscopic driven duffing oscillators *Phys. Rev.* E **84** 011144

[6] Dykman M I, Maloney C M, Smelyanskiy V N and Silverstein M 1998 Fluctuational phase-flip transitions in parametrically driven oscillators *Phys. Rev.* E **57** 5202–12

[7] Marthaler M and Dykman M I 2006 Switching via quantum activation: a parametrically modulated oscillator *Phys. Rev.* A **73** 042108

[8] Marthaler M and Dykman M I 2007 Quantum interference in the classically forbidden region: a parametric oscillator *Phys. Rev.* A **76** 010102

[9] Svensson I-M, Bengtsson A, Krantz P, Bylander J, Shumeiko V and Delsing P 2017 Period-tripling subharmonic oscillations in a driven superconducting resonator *Phys. Rev.* B **96** 174503

[10] Zhang Y, Gosner J, Girvin S M, Ankerhold J and Dykman M I 2017 Time-translation-symmetry breaking in a driven oscillator: from the quantum coherent to the incoherent regime *Phys. Rev.* A **96** 052124

[11] Zhang Y and Dykman M I 2019 Nonlocal random walk over Floquet states of a dissipative nonlinear oscillator *Phys. Rev.* E **100** 052148
[12] Lörch N, Zhang Y, Bruder C and Dykman M I 2019 Quantum state preparation for coupled period tripling oscillators *Phys. Rev. Res* **1** 023023
[13] Gosner J, Kubala B and Ankerhold J 2020 Relaxation dynamics and dissipative phase transition in quantum oscillators with period tripling *Phys. Rev.* E **101** 054501
[14] Svensson I-M, Bengtsson A, Bylander J, Shumeiko V and Delsing P 2018 Period multiplication in a parametrically driven superconducting resonator *Appl. Phys. Lett.* **113** 022602
[15] Buchleitner A, Delande D and Zakrzewski J 2002 Nondispersive wave packets in periodically driven quantum systems *Phys. Rep.* **368** 409–547
[16] Sacha K 2015 Modeling spontaneous breaking of time-translation symmetry *Phys. Rev.* A **91** 033617
[17] Giergiel K, Tran T, Zaheer A, Singh A, Sidorov A, Sacha K and Hannaford P 2020 Creating big time crystals with ultracold atoms *New J. Phys.* **22** 085004
[18] Sacha K 2020 *Discrete Time Crystals and Related Phenomena* (Cham: Springer) pp 39–172
[19] Guo L and Marthaler M 2016 Synthesizing lattice structures in phase space *New J. Phys.* **18** 023006
[20] Zaslavsky G M, Zakharov M Y, Sagdeev R Z, Usikov D A and Chernikov A A 1986 Usikov D A and Chernikov A A 1986 Stochastic web and diffusion of particles in a magnetic field *Zh. Eksp. Teor. Fiz.* **91** 500
[21] Zaslavsky G M 2008 *Hamiltonian Chaos and Fractional Dynamics* (Oxford: Oxford University Press)
[22] Berman G P, Rubaev V Y and Zaslavsky G M 1991 The problem of quantum chaos in a kicked harmonic oscillator *Nonlinearity* **4** 543–66
[23] Carvalho A R R and Buchleitner A 2004 Web-assisted tunneling in the kicked harmonic oscillator *Phys. Rev. Lett.* **93** 204101
[24] Liang P, Marthaler M and Guo L 2018 Floquet many-body engineering: topology and many-body physics in phase space lattices *New J. Phys.* **20** 023043
[25] Guo L and Liang P 2020 Condensed matter physics in time crystals *New J. Phys.* **22** 075003
[26] Guo L, Peano V and Marquardt F 2021 Phase space crystal vibrations: chiral edge states with preserved time-reversal symmetry arXiv:2105.06989 (submitted and under review, see the updated publication information)
[27] Landau L D and Lifshitz E M 1977 *Quantum Mechanics (Course of Theoretical Physics: v.3)* (Oxford: Pergamon)
[28] Kim K, Heo M-S, Lee K-H, Jang K, Noh H-R, Kim D and Jhe W 2006 Spontaneous symmetry breaking of population in a nonadiabatically driven atomic trap: an Ising-class phase transition *Phys. Rev. Lett.* **96** 150601
[29] Heo M-S, Kim Y, Kim K, Moon G, Lee J, Noh H-R, Dykman M I and Jhe W 2010 Ideal mean-field transition in a modulated cold atom system *Phys. Rev.* E **82** 031134
[30] Wustmann W and Shumeiko V 2019 Parametric effects in circuit quantum electrodynamics *Low Temp. Phys.* **45** 848–69
[31] Sacha K and Delande D 2016 Anderson localization in the time domain *Phys. Rev.* A **94** 023633
[32] Delande D, Morales-Molina L and Sacha K 2017 Three-dimensional localized-delocalized Anderson transition in the time domain *Phys. Rev. Lett.* **119** 230404
[33] Giergiel K and Sacha K 2017 Anderson localization of a rydberg electron along a classical orbit *Phys. Rev.* A **95** 063402

[34] Sacha K 2015 Anderson localization and Mott insulator phase in the time domain *Sci. Rep.* **5** 10787

[35] Sacha K and Zakrzewski J 2017 Time crystals: a review *Rep. Prog. Phys.* **81** 016401

[36] Bloch F 1929 Über die Quantenmechanik der Elektronen in Kristallgittern *Zeitschrift für Physik* **52** 555–600

[37] Grosso G and Parravicini G P 2000 I—Electrons in one-dimensional periodic potentials *Solid State Physics* (London: Academic) 1–36 978-0-12-304460-0 https://www.sciencedirect.com/science/article/pii/B9780123044600500013

[38] Pechal M, Berger S, Abdumalikov A A, Fink J M, Mlynek J A, Steffen L, Wallraff A and Filipp S 2012 Geometric phase and nonadiabatic effects in an electronic harmonic oscillator *Phys. Rev. Lett.* **108** 170401

[39] Hofstadter D R 1976 Energy levels and wave functions of Bloch electrons in rational and irrational magnetic fields *Phys. Rev.* B **14** 2239–49

[40] Harper P G 1955 The general motion of conduction electrons in a uniform magnetic field, with application to the diamagnetism of metals *Proc. Phys. Soc.* A **68** 879–92

[41] Zak J 1972 *The Kq-Representation in the Dynamics of Electrons in Solids* vol 27 (New York: Academic) pp 1–62

[42] Dana I 1995 Extended and localized states of generalized kicked harper models *Phys. Rev.* E **52** 466–72

[43] Prado Reynoso M Á, López Vázquez P C and Gorin T 2017 Quantum kicked harmonic oscillator in contact with a heat bath *Phys. Rev.* A **95** 022118

[44] Hone D W, Ketzmerick R and Kohn W 2009 Statistical mechanics of Floquet systems: the pervasive problem of near degeneracies *Phys. Rev.* E **79** 051129

[45] Ketzmerick R and Wustmann W 2010 Statistical mechanics of Floquet systems with regular and chaotic states *Phys. Rev.* E **82** 021114

[46] Shirai T, Thingna J, Mori T, Denisov S, Hänggi P and Miyashita S 2016 Effective Floquet-Gibbs states for dissipative quantum systems *New J. Phys.* **18** 053008

[47] Xiao D, Chang M-C and Niu Q 2010 Berry phase effects on electronic properties *Rev. Mod. Phys.* **82** 1959–2007

[48] Asbóth J K, Oroszlány L and Pályi A 2016 *Berry Phase, Chern Number* (Cham: Springer) pp 23–44

[49] Fukui T, Hatsugai Y and Suzuki H 2005 Chern numbers in discretized Brillouin zone: efficient method of computing (spin) hall conductances *J. Phys. Soc. Japan* **74** 1674–7

[50] Gerry C and Knight P 2004 *Introductory Quantum Optics* (Cambridge: Cambridge University Press)

[51] Gottesman D, Kitaev A and Preskill J 2001 Encoding a qubit in an oscillator *Phys. Rev.* A **64** 012310

[52] Steane A M 1997 Active stabilization, quantum computation, and quantum state synthesis *Phys. Rev. Lett.* **78** 2252–5

[53] Flühmann C, Nguyen T L, Marinelli M, Negnevitsky V, Mehta K and Home J P 2019 Encoding a qubit in a trapped-ion mechanical oscillator *Nature* **566** 513–7

[54] Campagne-Ibarcq P *et al* 2020 Quantum error correction of a qubit encoded in grid states of an oscillator *Nature* **584** 368–72

[55] Rymarz M, Bosco S, Ciani A and DiVincenzo D P 2021 Hardware-encoding grid states in a nonreciprocal superconducting circuit *Phys. Rev.* X **11** 011032

[56] Conrad J 2021 Twirling and Hamiltonian engineering via dynamical decoupling for Gottesman-Kitaev-Preskill quantum computing *Phys. Rev.* A **103** 022404

Chapter 4

Many-body physics in phase space

Phase space lattices discussed in the last chapter focus on the single-particle quantum dynamics. In this chapter, we consider many interacting particles subject to a periodic driving field. We discuss classical many-body dynamics in section 4.1 and quantum many-body effects in section 4.2. Both in the classical limit and quantum regime, we establish the relationship between the effective interaction in phase space, and the original interaction in real space. A key finding is that a short-range interaction (e.g., contact interaction) in real space becomes a long-range interaction (e.g., Coulomb-like form) in phase space. With the ingredients of phase space lattice and phase space interaction, many strongly correlated models can be simulated in phase space resulting in novel condensed matter in dynamical systems.

4.1 Classical many-body dynamics

4.1.1 Classical phase space interaction

We consider many identical particles confined in a 1D potential and driven by an external driving field with single-particle Hamiltonian $H^s(x, p, t)$. We assume the interaction potential between particles $V(x_i - x_j)$ is a function of two particles' distance $|x_i - x_j|$. The many-body Hamiltonian then is described by

$$H = \sum_i H^s(x_i, p_i, t) + \sum_{i<j} V(x_i - x_j). \tag{4.1}$$

We transform the above Hamiltonian into the rotating frame with frequency ω using the generating function of the second kind (see section 2.1.3)

$$G_2(\boldsymbol{x}, \boldsymbol{P}, t) = \sum_i \frac{x_i P_i}{\cos \omega t} - \frac{1}{2} x_i^2 \tan \omega t - \frac{1}{2} P_i^2 \tan \omega t. \tag{4.2}$$

Here, the vectors $\boldsymbol{x} = (x_1, x_2, \cdots)$ and $\boldsymbol{P} = (P_1, P_2, \cdots)$ represent the assemble canonical coordinates of all the particles. According to equation (2.47), we have the corresponding canonical transformation for phase space coordinates $p_i = \partial G_2/\partial x_i$, $X_i = \partial G_2/\partial P_i$, i.e.,

$$\begin{cases} x_i(t) &= P_i(t)\sin\omega t + X_i(t)\cos\omega t \\ p_i(t) &= P_i(t)\cos\omega t - X_i(t)\sin\omega t \end{cases} \tag{4.3}$$

and for the Hamiltonian $H(\boldsymbol{X}, \boldsymbol{P}, t) = H(\boldsymbol{x}, \boldsymbol{p}, t) + \partial G_2/\partial t$. By dropping fast oscillating terms in $H(\boldsymbol{X}, \boldsymbol{P}, t)$, we get the rotating-wave approximation (RWA) Hamiltonian of all interacting particles in the rotating frame,

$$H_{\text{RWA}} = \sum_i H^s_{\text{RWA}}(X_i, P_i) + \sum_{i<j} U(X_i, P_i; X_j, P_j). \tag{4.4}$$

Here, $H^s_{\text{RWA}}(X_i, P_i)$ is the single-particle RWA Hamiltonian discussed in the last chapter. The interaction term $U(X_i, P_i; X_j, P_j)$ is the RWA part of the transformed interaction potential, which is equivalent to the time-averaged real space interaction $V(x_i - x_j)$ from equation (4.3), i.e.,

$$\begin{aligned} U_{ij} &= \frac{\omega}{2\pi}\int_0^{2\pi/\omega} V(\Delta X_{ij}\cos\omega t + \Delta P_{ij}\sin\omega t)dt \\ &\equiv \frac{\omega}{2\pi}\int_0^{2\pi/\omega} V[R_{ij}\cos(\omega t - \theta_{ij})]dt. \end{aligned} \tag{4.5}$$

Here, the parameters R_{ij} and θ_{ij} are defined via $R_{ij}\exp(i\theta_{ij}) \equiv (\Delta X_{ij}, \Delta P_{ij})$ with the relative displacement of two particles in phase space $(\Delta X_{ij}, \Delta P_{ij}) \equiv (X_i - X_j, P_i - P_j)$. We call the new variable R_{ij} the *phase space distance*

$$R_{ij} \equiv \sqrt{(X_i - X_j)^2 + (P_i - P_j)^2}. \tag{4.6}$$

A more general form for the phase space distance is $R_{ij} = \sqrt{(X_i - X_j)^2 + g^2(P_1 - P_2)^2}$, where g is a parameter with dimension $[X]/[P]$ similar to the speed of light used to connect space and time in the theory of relativity. In our framework, g is an engineered parameter by scaling procedure. If $g \neq 1$ and $g \neq 0$, one can always reset it to unit value by the canonical transformation $P_i \to gP_i$ and $H_{\text{RWA}} \to gH_{\text{RWA}}$.

Given an interaction potential $V(x_i - x_j)$ in real space, we write it in the Fourier form, i.e., $V(x) = \int_{-\infty}^{+\infty} dq V_q e^{iqx}$. Using the Jacobi–Anger expansion, $e^{iz\cos\varphi} = \sum_{-\infty}^{+\infty} i^m J_m(z) e^{im\varphi}$ with $J_0(\bullet)$ the Bessel function of mth order, we have

$$\begin{aligned} U_{ij} &= \frac{\omega}{2\pi}\sum_{m=-\infty}^{+\infty} i^m \int_{-\infty}^{+\infty} dq V_q \int_0^{2\pi/\omega} J_m(qR_{ij})e^{im(\omega t - \theta_{ij})} \\ &\approx \int_{-\infty}^{+\infty} dq V_q J_0(qR_{ij}). \end{aligned} \tag{4.7}$$

Here in the last step, we have assumed the change of R_{ij} and θ_{ij} is slow compared to the rotating frequency ω. Physically, this means the motion of unperturbed Hamiltonian is close to the harmonic motion given by the canonical transformation (4.3). That is to say, the total motion of a particle can be separated into a fast global oscillation with frequency ω and a much slower motion representing the dynamics of the phase and amplitude of the global motion due to the driving field and particles' interactions. However, if the unperturbed motion is far away from the harmonic oscillation, e.g., the example of bouncing atom on an oscillating mirror (gravitational bouncer) introduced in section 3.1.1, it is better to choose the canonical transformation given by the phase flow of unperturbed motion, see equation (2.52). Then, the effective interaction is the time-averaged real space interaction

$$U_{ij} = \frac{2\pi}{\omega}\int_0^{2\pi/\omega} V[x_i(t) - x_j(t)]dt. \tag{4.8}$$

In general, the explicit form of U is a function of both coordinates and momenta of two particles in the canonically transformed frame, i.e., $U = U(X_i, P_i; X_j, P_j)$. From the integral representation of the Bessel function, i.e., $J_0(x) = (2\pi)^{-1}\int_{-\pi}^{\pi} \exp(-ix \sin\tau)d\tau$, we can rewrite equation (4.7) in an alternative form as following

$$U(R_{ij}) = \int_{-\infty}^{+\infty} dq V_q J_0(qR_{ij}) = \frac{1}{2\pi}\int_0^{2\pi} V(R_{ij}\sin\tau)d\tau, \tag{4.9}$$

which is a special case of equation (4.8) for the harmonic oscillation. Since $U(R_{ij})$ is defined in the phase space of the rotating frame and it is only a function of phase space distance R_{ij}, we call it the *phase space interaction potential*. Below, we will calculate explicitly the phase space interactions for some example interactions [1].

4.1.1.1 Example I: contact interaction

The point-like contact interaction is used to describe the effective interaction between neutral ultracold atoms from the s-wave scattering process [2]. The contact interaction is modelled by the δ function, i.e.,

$$V(x_i - x_j) = \varepsilon\delta(x_i - x_j) = \lim_{\sigma\to 0}\frac{\varepsilon}{\pi}\frac{\sigma}{(x_i - x_j)^2 + \sigma^2}. \tag{4.10}$$

We introduce the Lorentz function here, because it can be used for the numerical simulation. Here, ε is the strength of the δ function interaction. The Fourier transformation coefficient of the Lorentz function is $V_q = \varepsilon e^{-|q|\sigma}/(2\pi)$. One can obtain the phase space interaction potential from equation (4.7) or equation (4.9)

$$U(R_{ij}) = \frac{\varepsilon}{\pi}\frac{1}{\sqrt{R_{ij}^2 + \sigma^2}}. \tag{4.11}$$

In the limit of $\sigma \to 0$, we have the phase space interaction potential

$$U(R_{ij}) = \frac{\varepsilon}{\pi R_{ij}}. \tag{4.12}$$

It is interesting to find that the short-distance δ interaction potential produces an effective long-distance Coulomb-like interaction as a function of phase space distance R_{ij}. Phase space interaction describes the stroboscopic dynamics of system. Two particles always collide with each other during one stroboscopic time period, and the phase space interaction is the average of their real interaction during the stroboscopic time period.

4.1.1.2 Example II: repulsive power-law interaction

A more general interaction potential form is the repulsive power-law interaction potential,

$$V(x_i - x_j) = \varepsilon^{2n}/\left|x_i - x_j\right|^{2n}$$

with integer and half integer $n \geqslant 1/2$. If $n = 1/2$, the potential is the Coulomb potential. If $n \to \infty$, the potential becomes the hard-sphere potential with a diameter ε. However, one encounters divergence problem in calculating the phase space interaction potential $U(R_{ij})$ from equation (4.7) or equation (4.9). The origin of divergence comes from the fact that $V(x_i - x_j) \to \infty$ as $|x_i - x_j| \to 0$, which indicates the two particles can never touch each other due to the infinite potential barrier. In [1], the correct renormalised $U(R_{ij})$ was obtained by introducing a cutoff τ_c for the integral of equation (4.9), i.e.,

$$\tilde{U}(R_{ij}) = \frac{2}{\pi}\int_{\tau_c}^{\frac{\pi}{2}} V(R_{ij}\sin\tau)d\tau. \tag{4.13}$$

The smallest distance r_c the two particles can approach during collision depends on their relative kinetic energy. If the relative kinetic energy of two particles is much larger than their interaction energy, the collision distance is very short, i.e., $r_c \ll 1$. When the two particles are far away, the interaction energy can be neglected and their kinetic energy in the centre-of-mass frame is $E_{\text{kin.}}(R_{ij}) = \frac{1}{4}R_{ij}^2$. We can calculate r_c according to the energy conservation law in the centre-of-mass frame

$$V(r_c) = E_{\text{kin.}}(R_{ij}) = \frac{1}{4}R_{ij}^2. \tag{4.14}$$

It is important to note here that the phase space interaction potential $U(R_{ij})$ is in fact a weighted time average of the interaction in the laboratory frame as revealed by equation (4.9). Because of the repulsive power-law interaction, two particles already slow down when they start to collide during an oscillation period. Therefore, we connect τ_c and r_c by

$$r_c = \gamma R_{ij}\sin\tau_c \approx \gamma R_{ij}\tau_c \tag{4.15}$$

with the important *collision factor* $\gamma \gtrsim 1$, which is bigger than one, because of the extra interaction energy the particles accumulate as they move closer to each other and start to slow down. Therefore we integrate to a radius that can be smaller than r_c.

Combining equations (4.13), (4.14) and (4.15), we can obtain the explicit form of renormalised phase space interaction potential for power-law interaction [1]

$$\widetilde{U}(R_{ij}) = \begin{cases} \frac{2\varepsilon}{\pi R_{ij}} \ln(\varepsilon^{-1}\gamma R_{ij}^{3}/2), & \text{for } n = \frac{1}{2} \\ \frac{2\varepsilon\gamma^{2n-1}4^{\frac{1}{2n}-1}}{\pi(2n-1)} R_{ij}^{1-\frac{1}{n}}, & \text{for } n = 1, \frac{3}{2}, 2, \frac{5}{2}, \cdots \\ \frac{\varepsilon}{\pi} R_{ij}, & \text{for } n \to \infty. \end{cases} \tag{4.16}$$

Until now, the collision factor γ is still an unknown parameter phenomenologically introduced in equation (4.15). Comparing the above expression for $n = 1, \frac{3}{2}, 2, \frac{5}{2}, \cdots$ to the expression for $n \to \infty$, we get the correspondence condition

$$\gamma^{2n-1}/(2n-1) \to 2 \quad \text{for} \quad n \to \infty. \tag{4.17}$$

The simplest assumption is $\gamma^{2n-1} = 4n + c$, where c is a free parameter to be further determined. Thus the collision factor takes the form of $\gamma = (4n + c)^{1/(2n-1)}$. This form needs to be valid for the Coulomb potential, i.e., $n = 1/2$. By writing $n = 1/2 + \varepsilon/4$ with $\varepsilon \to 0$, we have

$$\gamma = (2 + c + \varepsilon)^{2/\varepsilon} \to (2 + c)^{2/\varepsilon} e^{2(2+c)} \quad \text{for} \quad \varepsilon \to 0.$$

The parameter c can only take the value of -1 to get a meaningful result. Otherwise, the prefactor $(2 + c)^{2/\varepsilon} \to (2 + c)^{\infty}$ takes either zero or infinity, both of which are unphysical. Finally, we get the collision factor γ for the inverse power-law interaction potential

$$\gamma = (4n - 1)^{\frac{1}{2n-1}}. \tag{4.18}$$

The collision factor becomes $\gamma = e^2$ for Coulomb interaction ($n = 1/2$) and $\gamma = 1$ for hard-sphere interaction ($n \to \infty$). In figure 4.1, we plot the real space power-law interaction potential and the corresponding phase space interaction potential. Several interesting features of phase space interaction should be pointed out as follows.

- For the Coulomb potential ($n = 1/2$), $U(R_{ij})$ still keeps the form of Coulomb's law, up to logarithmic corrections.
- For the special case of $n = 1$, $U(R_{ij})$ is a constant meaning there is no effective interaction in phase space.

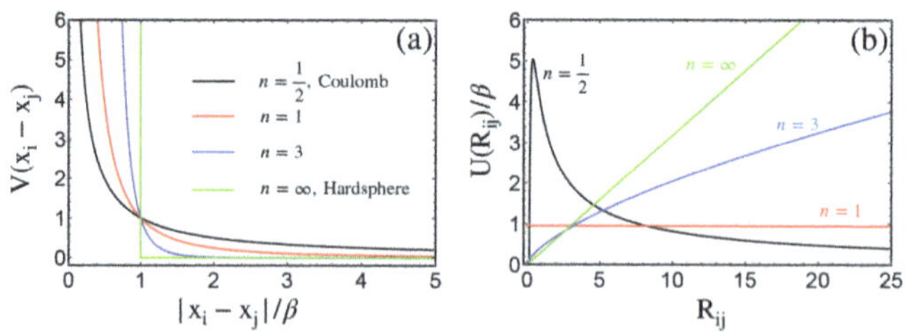

Figure 4.1. Classical phase space interaction potentials. (a) Real space power-law interaction potentials. The coloured dashed curves are the example plots for $n = 1/2$ (Coulomb interaction), $n = 1$, $n = 3$ and $n = \infty$ (hard-sphere interaction). (b) Phase space interaction potentials corresponding to the power-law potentials shown in figure (a). Parameter: $\varepsilon = 0.01$. Produced using own data but following methods in [1].

- For the special case of $n > 1$, $U(R_{ij})$ even grows with R_{ij}.
- For the hard-sphere interaction ($n \to \infty$), $U(R_{ij})$ increases linearly with phase space distance, reminiscent of the confinement interaction between quarks in quantum chromodynamics (QCD).

4.1.1.3 Example III: Lennard–Jones interaction potential

Another general choice to describe the interaction between two noble atoms or molecules is the Lennard–Jones interaction potential

$$V(x_i - x_j) = 4\epsilon\left(\frac{\sigma^{2m}}{|x_i - x_j|^{2m}} - \frac{\sigma^m}{|x_i - x_j|^m}\right), \tag{4.19}$$

where ϵ defines the interaction strength, σ defines the interaction range and the parameter m is usually taken $m = 6$ in the study. By introducing the cutoff τ_c, we obtain the phase space interaction potential from equation (4.13)

$$\tilde{U}(R_{ij}) = \frac{8\epsilon}{\pi}\left(\frac{\sigma^{2m}}{R_{ij}^{2m}}\frac{1}{2m-1}\tau_c^{1-2m} - \frac{\sigma^{2m}}{R_{ij}^{m}}\frac{1}{m-1}\tau_c^{1-m}\right). \tag{4.20}$$

Then from equation (4.14) and equation (4.15) we can calculate the cutoff as following

$$\tau_c = \frac{\sigma}{\gamma R_{ij}}\left(\frac{1}{2} + \frac{1}{2}\sqrt{1 + \frac{1}{4\epsilon}R_{ij}^2}\right)^{-1/m}. \tag{4.21}$$

As seen from equation (4.19), the Lennard–Jones potential is composed of two inverse power-law potential with exponents $2m$ and m respectively. During the collision in the range of small distance $|x_i - x_j| < \sigma$, we have $\frac{\sigma^{2m}}{|x_i - x_j|^{2m}} \gg \frac{\sigma^m}{|x_i - x_j|^m}$. Therefore, the term $\frac{\sigma^{2m}}{|x_i - x_j|^{2m}}$ in equation (4.19) is dominant during the collision. The collision factor of the Lennard–Jones potential can be calculated from equation (4.18) by choosing $n = m$, i.e., $\gamma = (4m - 1)^{\frac{1}{2m-1}}$.

4.1.2 Newton's laws in phase space

For the system of many interacting particles in 1D real space, we can label each particle in the 2D phase space by the position vector in phase space $\boldsymbol{Z}_i \equiv (X_i, P_i)^T$. We define a unit normal vector $\hat{\boldsymbol{n}}$ perpendicular to the 2D phase space plane, i.e., $\hat{\boldsymbol{n}} \equiv \hat{\boldsymbol{n}}_X \times \hat{\boldsymbol{n}}_P$, where $\hat{\boldsymbol{n}}_X$ and $\hat{\boldsymbol{n}}_P$ are the unit vectors in X and P direction respectively. Note that the phase space interaction potential depends on the relative momentum of two particles via the phase space distance R_{ij}. It is nature to extend the concept of force from real space to phase space, and define the *phase space force* for the ith particle as following

$$\boldsymbol{F}_i \equiv -\nabla_i H = -\left(\frac{\partial H}{\partial X_i}, \frac{\partial H}{\partial P_i}\right)^T. \tag{4.22}$$

Using the phase space force, we can rewrite the canonical equation of motion, see equation (2.14), in the following compact form

$$\frac{d}{dt}\boldsymbol{Z}_i = \hat{\boldsymbol{n}} \times \boldsymbol{F}_i. \tag{4.23}$$

It is interesting to find that a phase space force causes a displacement of particle *perpendicular* to the force direction as illustrated by the right panel in figure 1.1.

For the two-body phase space interaction potential $U(R_{ij})$, the phase space force on the ith particle exerted by the jth particle is

$$\boldsymbol{F}_{ij} \equiv -\nabla_i U(R_{ij}) = U'(R_{ij})\hat{\boldsymbol{n}}_{ij}, \text{ with } \hat{\boldsymbol{n}}_{ij} \equiv \frac{1}{R_{ij}}\left(X_j - X_i, P_j - P_i\right)^T. \tag{4.24}$$

Here, $\hat{\boldsymbol{n}}_{ij}$ is the unit direction vector of the force exerted by the second particle upon the first particle. Similarly, the force on the second particle is $\boldsymbol{F}_{ji} = -\boldsymbol{F}_{ij}$. Based on equations (4.23) and (4.24), we have the Newton's laws of motion in 2D phase space as follows.

1. *In phase space, an object remains at rest unless acted upon by a phase space force.*
2. *In phase space, the total phase space force on an object is proportional to and perpendicular to the object's phase space velocity:* $\boldsymbol{F}_i = \boldsymbol{V}_i \times \hat{\boldsymbol{n}}$.
3. *In phase space, the interaction forces between two objects have equal magnitude and opposite direction.*

In the second law, we have used the vector calculus identity $\boldsymbol{A} \times (\boldsymbol{B} \times \boldsymbol{C}) = \boldsymbol{B}(\boldsymbol{A} \cdot \boldsymbol{C}) - \boldsymbol{C}(\boldsymbol{A} \cdot \boldsymbol{B})$ and rewrite equation (4.23) in an equivalent form $\boldsymbol{F}_i = \boldsymbol{V}_i \times \hat{\boldsymbol{n}}$, where $\boldsymbol{V}_i \equiv d\boldsymbol{Z}_i/dt$ is defined as the *phase space velocity*. We emphasise that the 'forces' mentioned above all refer to the phase space forces not the real space forces. For example, a free particle in real space with Hamiltonian $H = P^2/2$ is actually exerted by a phase space force $\boldsymbol{F} = -P\hat{\boldsymbol{n}}_P$. As a result, the particle has a constant phase space velocity $d\boldsymbol{Z}/dt = P\hat{\boldsymbol{n}}_X$ from the equation of motion (EOM) (4.23), i.e., the particle is moving uniformly forward along the spatial direction $\hat{\boldsymbol{n}}_X$ reflecting the principle of inertia.

Phase space force has the same form as Lorentz force, i.e., $\boldsymbol{F}_L = q\boldsymbol{v} \times \boldsymbol{B} = qB\boldsymbol{v} \times \hat{\boldsymbol{n}}$, where $\boldsymbol{B} = B\hat{\boldsymbol{n}}$ is the magnetic field and q is the charge of particle. In real space, the dynamics is determined by the Lorentz equation, i.e., $m\frac{d}{dt}\boldsymbol{v} = q(\boldsymbol{E} + \boldsymbol{v} \times \boldsymbol{B})$. In the limit of infinitely large charge-to-mass ratio $q/m \to \infty$, the Lorentz equation of motion becomes $\boldsymbol{v}_i = (qB)^{-1}\boldsymbol{F}_i \times \hat{\boldsymbol{n}}$, which has the same form as the phase space dynamics, see equation (4.23).

Magnetron motion We consider the motion of a particle with mass m and charge q in a 2D x–y plane. There is a uniform magnetic field $\boldsymbol{B} = B\hat{\boldsymbol{n}}$ perpendicular to the x–y plane and a uniform electric field $\boldsymbol{E} = -\nabla\Phi = (E_x, E_y)$ in the 2D plane. Then, the motion of particle is given by the Lorentz equation

$$m\frac{d\boldsymbol{v}}{dt} = q(\boldsymbol{E} + \boldsymbol{v} \times \mathbf{B}), \tag{4.25}$$

where $\boldsymbol{v} = d\boldsymbol{r}/dt = (v_x, v_y)$ is the velocity of particle and $\boldsymbol{r}(t) = (x, y)$ is the position of particle. The general solution of EOM (4.25) is given by

$$\begin{cases} \dfrac{dx}{dt} = \dfrac{E_y}{B} + \omega_c r_c \cos(\omega_c t + \phi) \\ \dfrac{dy}{dt} = -\dfrac{E_x}{B} - \omega_c r_c \sin(\omega_c t + \phi). \end{cases} \tag{4.26}$$

Here, $\omega_c = \frac{qB}{m}$ is the cyclotron frequency and r_c is the cyclotron radius given by

$$r_c = \frac{1}{\omega_c}\sqrt{\left(\frac{E_y}{B} - v_x^0\right)^2 + \left(\frac{E_x}{B} + v_y^0\right)^2}.$$

The parameters r_c and ϕ are determined by the initial condition of velocity (v_x^0, v_y^0). The total motion of a charged particle is composed of the *cyclotron motion* and the *magnetron motion* describing the drift of the cyclotron motion centre with drift velocity $(E_y/B, -E_x/B)$. In the limit of infinity large charge-to-mass ratio $q/m \to \infty$, the cyclotron radius becomes zero $r_c \to 0$, and the dynamics of magnetron motion can be obtained from equation (4.26)

$$\frac{d\boldsymbol{r}}{dt} = \hat{\boldsymbol{n}} \times (-B^{-1}\boldsymbol{E}). \tag{4.27}$$

Now, $\boldsymbol{r}$ describes the displacement of magnetron motion, i.e., the drifted centre of cyclotron motion. Note that we have assumed a uniform $\boldsymbol{E}$ in the above discussion. In fact, the magnetron EOM (4.27) can be extended to arbitrary electric field $\boldsymbol{E}$ if the condition $r_c \to 0$ is always satisfied, i.e., the cyclotron radius r_c is neglectable during the motion. If we map the 2D x–y plane to the 2D phase space plane and take $\boldsymbol{E} = B\nabla H$, namely, taking the electric potential $\Phi = -BH$, the magnetron EOM (4.27) has the same form of phase space EOM given by equation (4.23).

4.1.2.1 Two-body problem

As a concrete example, we consider two particles trapped in the same harmonic potential with interaction potential $V(x_1 - x_2)$

$$H = \sum_{i=1,2} \frac{p_i^2}{2} + \frac{x_i^2}{2} + V(x_1 - x_2). \tag{4.28}$$

By choosing rotating frequency $\omega = 1$ in the canonical transformation (4.2), we have the RWA Hamiltonian in the rotating frame

$$H_{\mathrm{RWA}} = U(R_{12}), \tag{4.29}$$

which is purely determined by the phase space interaction potential of two particles. If there is no interaction, the two particles stay at two fixed points in phase space. This is obvious because the rotating frame has the same frequency as the harmonic oscillation. For finite interaction, the two particles will start to move in phase space. Below, we will solve the motion of two particles using the Newton's laws in phase space.

We first define the vector $\hat{\boldsymbol{n}}_{12}$ as the unit vector directing from the first particle to the second particle. We then introduce the vector $\boldsymbol{Z}_0 \equiv (\boldsymbol{Z}_1 + \boldsymbol{Z}_2)/2$ which describes the motion of two particles' centre. According to the 2nd and 3rd laws, we have

$$\frac{d}{dt}\boldsymbol{Z}_0 = \hat{\boldsymbol{n}} \times \frac{1}{2}(\boldsymbol{F}_{12} + \boldsymbol{F}_{21}) = 0,$$

i.e., $\boldsymbol{Z}_0$ is a constant. The EOM of each particle is

$$\frac{d}{dt}(\boldsymbol{Z}_1 - \boldsymbol{Z}_0) = \hat{\boldsymbol{n}} \times \boldsymbol{F}_{12}, \Rightarrow \frac{d}{dt}(\boldsymbol{Z}_1 - \boldsymbol{Z}_2) = -2U'(R_{12})\hat{\boldsymbol{n}} \times \hat{\boldsymbol{n}}_{12}. \tag{4.30}$$

Dot produced by the vector $(\boldsymbol{Z}_1 - \boldsymbol{Z}_2)$ on both sides of the above equation, we have

$$(\boldsymbol{Z}_1 - \boldsymbol{Z}_2) \cdot \frac{d}{dt}(\boldsymbol{Z}_1 - \boldsymbol{Z}_2) = -2U'(R_{12})(\boldsymbol{Z}_1 - \boldsymbol{Z}_2) \cdot (\hat{\boldsymbol{n}} \times \hat{\boldsymbol{n}}_{12}) = 0. \tag{4.31}$$

Therefore, the phase space distance of two particles $R_{12} = |\boldsymbol{Z}_1 - \boldsymbol{Z}_2|$ is a constant. The EOM (4.30) is further simplified by dividing both sides with R_{12}

$$\frac{d}{dt}\hat{\boldsymbol{n}}_{12} = -\frac{2}{R_{12}}\frac{dU}{dR_{12}}\hat{\boldsymbol{n}} \times \hat{\boldsymbol{n}}_{12}. \tag{4.32}$$

The above equation means the unit vector $\hat{\boldsymbol{n}}_{12}$ is rotating with a constant angular velocity

$$\omega_R = -\frac{2}{R_{12}}\frac{dU}{dR_{12}} \tag{4.33}$$

and the rotating direction is given by $\hat{\boldsymbol{n}} \times \hat{\boldsymbol{n}}_{12}$. Analytical expressions of ω_R can be obtained from equation (4.33) and $U(R_{ij})$. For example, the rotating frequency is $\omega_R = 2\varepsilon/\pi R_{12}^3$ for the contact interaction (4.12).

The two-body rotation frequency ω_R can also be obtained from Poincaré mapping, i.e., we simulate the dynamics of two particles $(x_i(t), p_i(t))$ based on the original Hamiltonian (4.28) and take their values stroboscopically every time period $\Delta t = 2\pi$. This gives the trajectories of two particles in phase space as shown by the two dotted lines in figure 4.2(a). In figure 4.2(b), we compare the numerically extracted ω_R to the analytical expression equation (4.33) for different types of interactions, i.e., Lorenz interaction, Coulomb interaction, hard-sphere interaction and Lennard–Jones interaction. We see that the result (4.33) is very good for large R_{12} and breaks down for small R_{12}. From equation (4.15), we see the relative kinetic energy of two particles is $R_{12}^2/4$. Therefore, we conclude that *the frequency ω_R as a function of R_{12} given by equation (4.33) is valid when the interaction energy is much*

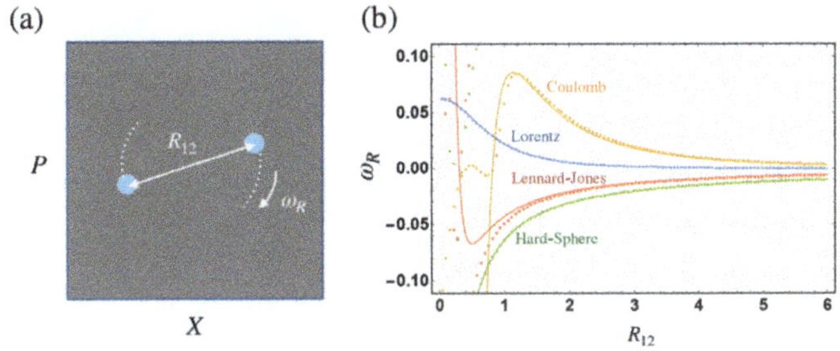

Figure 4.2. Two-body phase space dynamics. (a) The two solid blue dots represent the initial positions of two particles on phase space, which are fixed points without interaction. In the presence of interaction, the two particles start to rotate with a frequency ω_R, which depends on their phase space distance R_{12}. (b) Rotating frequency ω_R as function of R_{12}. The coloured dots represent the data from Poincaré mapping while the solid lines are given by our analytical result (4.33). Interaction potential parameters: $\varepsilon = 0.1$, $\sigma = 1$ for Lorenz interaction; $\varepsilon = 0.1$ for Coulomb interaction; $\epsilon = 0.01$, $\sigma = 0.1$ for Lennard–Jones interaction. Hard-sphere interaction is modelled by the inverse power-law interaction with $\varepsilon = 0.1$, $n = 20$. Produced using own data but following methods in [1].

smaller than the relative kinetic energy of two particles, i.e., the cutoff given by equation (4.15) satisfies $r_c \ll 1$.

4.1.2.2 Three-body dynamics

For three particles trapped in a harmonic potential with interaction potential $V(x_i - x_j)$, the many-body Hamiltonian is

$$H = \sum_{i=1,2,3} \frac{p_i^2}{2} + \frac{x_i^2}{2} + \sum_{i<j=1,2,3} V(x_i - x_j). \tag{4.34}$$

By choosing rotating frequency $\omega = 1$ in the canonical transformation (4.2), we have the phase space Hamiltonian in the rotating frame

$$H_{\text{RWA}} = U(R_{12}) + U(R_{13}) + U(R_{23}). \tag{4.35}$$

Not like the two-body dynamics, which can be solved analytically, the three-body dynamics cannot be obtained analytically in general. We thus solve the three-body problem via numerical simulations.

In figure 4.3, we show the numerical results for the Coulomb, hard-sphere and Lennard–Jones interactions. The coloured dots (blue, green and red) represent the three particles' initial positions in phase space, and the corresponding coloured solid lines are the trajectories obtained from the phase space Hamiltonian (4.35). The black dotted lines are the results from the original Hamiltonian (4.34) combined with Poincaré mapping. We see that the agreement is good with the discrepancy in the long time limit from the RWA. Depending on the type of interactions and the initial conditions, the three interacting particles may have complex trajectories in phase space. For Coulomb interaction, particle 2 and particle 3 can be viewed as a two-body subsystem while their centre of mass forms another larger two-body system with particle 1. Since the phase space interaction of Coulomb interaction decays with phase space distance, the rotational frequency of particle 1 around the

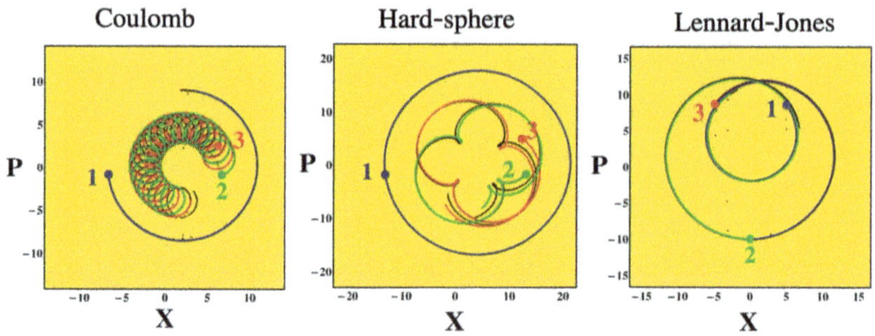

Figure 4.3. Three-body phase space dynamics. Trajectories of three interacting particles on phase space under Coulomb interaction with $\varepsilon = 0.1$ (left), hard-sphere interaction with $\varepsilon = 0.1$ (middle) and Lennard–Jones interaction with parameters $\epsilon = 0.01$ and $\sigma = 0.1$ (right). The coloured dots marked by numbers 1, 2 and 3 represent the initial conditions. The coloured solid curves represent the dynamics from phase space Hamiltonian (4.35). The black dotted curves are the data from the original Hamiltonian (4.34) by Poincaré mapping. Produced using own data but following methods in [1].

centre of particle 2 and 3 is much slower than the rotating frequency of particle 2 and 3 around their centre. For the hard-sphere interaction, the corresponding phase space interaction increases linearly with the phase space distance, which means the interaction between particle 1 and particle 2 or between particle 1 and particle 3 is stronger than interaction between particle 2 and particle 3. As a result, the orbits of particle 2 and particle 3 are not as regular as in the case of Coulomb interaction due to the strong disturbance by particle 1. For the Lennard–Jones interaction, we change the initial conditions of the three particles such that the three particles exchange their positions in phase space with a time period.

4.1.3 1D phase space crystals

The two examples discussed in section 4.1.2 are subjected to a static harmonic trap, and thus there is no background potential in phase space. Here, we introduce a background lattice potential in phase space via the driving field, and study the many-body dynamics from both the phase space lattice and the phase space interaction. We consider many interacting kicked harmonic oscillators with many-body Hamiltonian given by

$$H = \sum_i \frac{1}{2}(x_i^2 + p_i^2) + K\tau \cos x_i \sum_{n=-\infty}^{\infty} \delta(t - n\tau) + \sum_{i<j} V(x_i - x_j). \tag{4.36}$$

By choosing $2\pi/\tau = 4$ and using the canonical transformation (4.3), we have many interacting particles in a square lattice in phase space (see section 3.1.2). The total many-body phase space Hamiltonian given by

$$H = \sum_i \frac{1}{2}K(\cos X_i + \cos P_i) + \sum_{i<j} U(R_{ij}). \tag{4.37}$$

The explicit forms of phase space interaction $U(R_{ij})$ depend on the choice of real space interaction potential $V(x_i - x_j)$. Below, we discuss two typical cases of contact interaction and hard-sphere interaction.

In figure 4.4, we show the dynamics of seven interacting atoms with real space contact interaction potential $V(x_i - x_j) = \varepsilon\delta(x_i - x_j)$, which is modelled by a Lorentz function $V(x_i - x_j) = \frac{\varepsilon}{\pi}\frac{\sigma}{(x_i - x_j)^2 + \sigma^2}$ with $\varepsilon = 0.194$ and $\sigma = 0.1$ in our numerical simulations. The corresponding phase space interaction potential is given by $U(R_{ij}) = \frac{\varepsilon}{\pi}\frac{1}{R_{ij}}$. The seven particles are located initially at the lattices sites of phase space lattice as shown by the seven red dots. Due to the phase space interaction, the particles will oscillate around their equilibrium points in phase space when their interaction is weak. The two particles at the ends oscillate with the largest amplitude. If the interaction is strong enough, the two edge atoms can overcome the potential barrier of the phase space lattice, and destroy the crystal state. The existence of such crystal state is guaranteed by the condition that the oscillating amplitude of the edge atom $\Delta\mathbf{Z}_{\text{edge}}$ is smaller than the radius of the dashed circle indicated in figure 4.4, i.e., $\Delta\mathbf{Z}_{\text{edge}} < (\sqrt{2} - 1)\pi$. Using the linear expansion around the equilibrium points, one can estimate the oscillating amplitude of the edge atom [3], which is $\Delta\mathbf{Z}_{\text{edge}} = \frac{\varepsilon}{12\pi K}$ for an infinite number of atoms. Therefore, we conclude that the 1D crystal state in phase space exists if the interaction satisfies $|\varepsilon| < 12(\sqrt{2} - 1)\pi^2 K$.

In our numerical simulations, the hard-sphere interaction is modelled approximately by an inverse power-law potential $V(x_i - x_j) = \frac{\varepsilon^{2n}}{(x_i - x_j)^{2n}}$ with a high power $n = 10$. The corresponding phase space interaction potential is given by $U(R_{ij}) = \varepsilon\pi^{-1}R_{ij}$. The linear analysis gives the oscillating amplitudes of the two edge particles $\Delta\mathbf{Z}_{\text{edge}} = \frac{2\varepsilon}{\pi K}(N - 1)$, where N is the number of atoms. Different from the contact interaction, the oscillating amplitude under hard-sphere interaction $\Delta\mathbf{Z}_{\text{edge}}$ increases linearly with the number of atoms. Therefore, it is impossible to create an infinitely long 1D crystal state with hard-sphere interaction. For a given kicking strength K and hard-sphere radius ε, the critical atom number for the existence of 1D crystal state is $N_c \approx (\sqrt{2} - 1)\pi^2\frac{K}{2\varepsilon}$.

We call the crystal state in phase space formed by many atoms, the *phase space crystal*. One should distinguish the concepts of phase space lattice discussed in

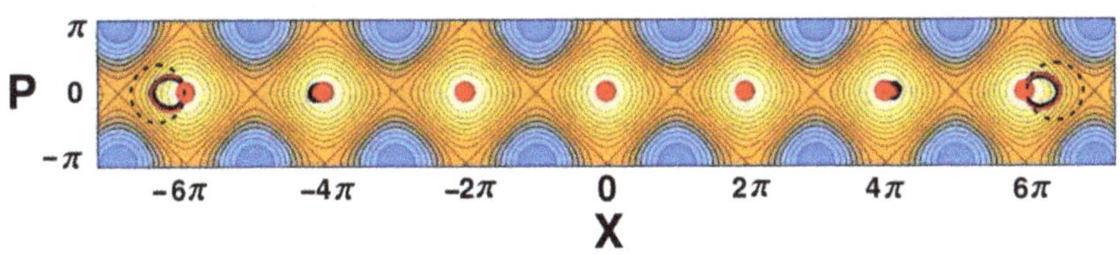

Figure 4.4. 1D phase space crystal. The initial positions of seven atoms are marked by red dots. Black lines and red lines are the phase space trajectories calculated from Poincaré map and phase space Hamiltonian (4.37) with contact interaction. The 1D phase space crystal survives if the two atoms at the ends oscillate inside the dashed circles. Parameters: kicking strength $K = -0.02\pi^{-1}$, interaction strength $\varepsilon = 0.194$. Reprinted from [4].

chapter 3 and the phase space crystal introduced here. Phase space lattice refers to the periodic structure in phase space of a single-particle Hamiltonian without consideration of atomic interaction, while the phase space crystal refers to the many-body state formed by interacting atoms. In the experiments, the 1D phase space crystal shown in figure 4.4 can be realised by two basic steps: first, prepare the initial state of atoms via applying a very strong static optical lattice; then, suddenly turn off the strong static optical lattice and add a weak optical lattice stroboscopically.

The atoms can also occupy the lattice sites both in X and P directions and form a stable crystalline state in phase space, i.e., the 2D phase space crystal. The existence condition and the linear dynamics of 2D phase space crystals around the equilibrium configuration is the central topic of chapter 5.

4.2 Quantum many-body effects

4.2.1 Phase space distance operator

In quantum mechanics, the phase space interaction introduced above in the classical regime becomes subtle since the phase space is a noncommutative space. As discussed in figure 3.6, the concept of *point* is meaningless in the non-commutative phase space. Instead, it is only allowed to define the coherent state $|\alpha\rangle$ as a '*point*' in noncommutative geometry. Similarly, the concept of *distance* also needs to be reexamined. It is natural to define the *phase space distance operator* by [4]

$$\hat{R}_{ij} \equiv \sqrt{(\hat{X}_i - \hat{X}_j)^2 + (\hat{P}_i - \hat{P}_j)^2}. \tag{4.38}$$

For two arbitrary particles, we introduce the operators $\hat{X}_c \equiv (\hat{X}_1 + \hat{X}_2)/2$, $\hat{P}_c \equiv (\hat{P}_1 + \hat{P}_2)/2$ representing the coordinator and momentum of two particles' centre of mass, and the operators $\Delta\hat{X} \equiv \hat{X}_1 - \hat{X}_2$, $\Delta\hat{P} \equiv \hat{P}_1 - \hat{P}_2$ representing their relative displacement in phase space. The phase space distance operator is $\hat{R} \equiv \sqrt{\Delta\hat{X}^2 + \Delta\hat{P}^2}$. From the commutation relationship $[\hat{X}_i, \hat{P}_j] = i\lambda\delta_{ij}$, we have $[\Delta\hat{X}, \Delta\hat{P}] = i(2\lambda)$, and $[\hat{R}^2, \hat{X}_c] = 0$, which means the motion of two particles' centre of mass and their relative motion are independent. Thus, we write the common eigenstate of commutative operators $\hat{R}^2$ and $\hat{X}_c$ as a product state $\Psi(X_1, X_2) = f(X_c)\Phi(X_1 - X_2)$, where the wave function $f(X_c)$ is the state of two particles' centre of mass and the wave function $\Phi(X_1 - X_2)$ describes their relative motion. Reminiscent of the Hamiltonian operator of a harmonic oscillator, the eigenvalues of operator $\hat{R}^2$ are given by $4\lambda(N + 1/2)$ with $N = 0, 1, 2, \cdots$. Therefore, the eigenvalues of the operator $\hat{R}$ are given by

$$R_N = 2\sqrt{\lambda(N + \frac{1}{2})}, \quad N = 0, 1, 2, \cdots. \tag{4.39}$$

For each N, the corresponding eigenstate is given by

$$\Phi_N(\Delta X) = \left(\frac{1}{\sqrt{2\lambda\pi}2^N N!}\right)^{\frac{1}{2}} H_N\left(\sqrt{\frac{1}{2\lambda}}\Delta X\right) e^{-\frac{1}{4\lambda}\Delta X^2}, \tag{4.40}$$

where $H_N(\bullet)$ is the Hermite polynomial of degree N. We choose functions $\delta(X_c - C)$, i.e., the eigenstate of operator $\hat{X}_c$, as the basis of two particles' centre of mass. Therefore, we use the Dirac notation $|N, C\rangle$ to represent the total eigenstate, which is determined by two good quantum numbers C and R_N, i.e., $\hat{X}_c|N, C\rangle = C|N, C\rangle$ and $\hat{R}|N, C\rangle = R_N|N, C\rangle$. In the coordinate representation, the total eigenstate has the explicit form [4]

$$\langle X_1, X_2|N, C\rangle = \Psi_{N,C}(X_1, X_2) = \delta(X_c - C)\Phi_N(\Delta X).$$

The distance of two particles in real space is a continuous variable from zero to infinity. However, the distance in phase space is a quantised variable and has a lower limit $\sqrt{2\lambda}$ as seen from equation (4.39). Here, we actually provide a description for the quantisation of the noncommutative background.

4.2.2 Quantum phase space interaction

In this section, we will discuss the quantum dynamics of interacting particles. The canonical transformation (4.3) in classical mechanics corresponds to the free time-evolution operator $\hat{O}(t) = \exp(i\lambda\omega\sum_i \hat{a}_i^\dagger \hat{a}_i t)$ in quantum mechanics, where we have defined the ladder operator $\hat{a}_i \equiv (\hat{x}_i + i\hat{p}_i)/\sqrt{2\lambda}$ with $[\hat{x}_i, \hat{p}_i] = i\lambda$. The transformation for the coordinates and momenta of particles are given by

$$\begin{cases} \hat{O}(t)\hat{x}_i\hat{O}^\dagger(t) &= \hat{P}_i \sin\omega t + \hat{X}_i \cos\omega t \\ \hat{O}(t)\hat{p}_i\hat{O}^\dagger(t) &= \hat{P}_i \cos\omega t - \hat{X}_i \sin\omega t. \end{cases} \tag{4.41}$$

Here, the operators $\hat{X}_i$ and $\hat{P}_i$ describe the dynamics of the ith atom's phase and amplitude. For the harmonic oscillator, $\hat{X}_i$ and $\hat{P}_i$ are fixed and correspond to the initial state of $\hat{x}_i(0)$ and $\hat{p}_i(0)$. If the time evolution of $\hat{X}_i$ and $\hat{P}_i$ is slow compared to the fast global harmonic oscillation, they can be obtained stroboscopically from the time evolution of $\hat{x}_i(t)$ and $\hat{p}_i(t)$ every time period of $2\pi/\omega$. The total Hamiltonian in the rotating frame is given by the canonical transformation, i.e., $\hat{O}(t)H(t)\hat{O}^\dagger(t) - iO(t)\dot{O}^\dagger(t)$. In the RWA, we drop the fast oscillating terms and arrive at the time-independent Hamiltonian [4]

$$H_{\text{RWA}} = \sum_i H_{\text{RWA}}(\hat{X}_i, \hat{P}_i) + \sum_{i<j} U(\hat{X}_i, \hat{P}_i; \hat{X}_j, \hat{P}_j). \tag{4.42}$$

Here, $H_{\text{RWA}}(\hat{X}_i, \hat{P}_i)$ is the single-particle RWA Hamiltonian, and $U(\hat{X}_i, \hat{P}_i; \hat{X}_j, \hat{P}_j)$ is the time-independent part of transformed real space interaction potential $\hat{O}(t)V(\hat{x}_i - \hat{x}_j)\hat{O}^\dagger(t)$. In this section, we aim to determine the explicit form of $U(\hat{X}_i, \hat{P}_i; \hat{X}_j, \hat{P}_j)$ in quantum mechanics.

4.2.2.1 General framework

In order to determine the phase space interaction potential $U(\hat{X}_1, \hat{P}_1; \hat{X}_2, \hat{P}_2)$, we transform the relative displacement of two particles in the rotating frame from the transformation (4.41), i.e., $\hat{O}(t)(\hat{x}_1 - \hat{x}_2)\hat{O}^\dagger(t) = \Delta\hat{P}\sin t + \Delta\hat{X}\cos t$. Therefore, for a given real space interaction potential, we have $\hat{O}V(\hat{x}_1 - \hat{x}_2)\hat{O}^\dagger = \int_{-\infty}^{+\infty} dq V_q \hat{Q}$ with the Fourier coefficients $V_q = \frac{1}{2\pi}\int_{-\infty}^{+\infty} dx V(x)e^{-iqx}$ and the operator $\hat{Q} \equiv \exp[iq(\Delta\hat{P}\sin\omega t + \Delta\hat{X}\cos\omega t)]$. The matrix element of the operator $\hat{Q}$ in the $\hat{R}$-representation discussed above is given by the Laguerre polynomials [3, 5]

$$\langle N|\hat{Q}|M\rangle = e^{-\frac{\lambda q^2}{2}+i(M-N)(\frac{\pi}{2}-\omega t)}\sqrt{\frac{N!}{M!}}(\lambda q^2)^{\frac{M-N}{2}}L_N^{M-N}(\lambda q^2), \tag{4.43}$$

where $|N\rangle$ and $|M\rangle$ are the eigenstates of the operator $\hat{R}$ given by equation (4.40). In the RWA, we only keep the time-independent diagonal elements of the matrix (4.43), i.e., $\langle N|\hat{Q}|N\rangle$ with $N = 0, 1, 2, \cdots$. Thus, given an arbitrary real space interaction potential $V(x_1 - x_2)$, we find a compact expression for the phase space interaction potential [4]

$$U(\hat{R}) = \int_{-\infty}^{+\infty} dq V_q e^{-\frac{\lambda q^2}{2}} L_{\frac{1}{4\lambda}\hat{R}^2 - \frac{1}{2}}(\lambda q^2). \tag{4.44}$$

In the eigenbasis $|N, C\rangle$, we have $U(\hat{R}) = \sum_N U(R_N)\int dC|N, C\rangle\langle N, C|$. Here, the interaction potential $U(\hat{R})$ takes the value $U(R_N)$ with $R_N = 2\sqrt{\lambda(N + 1/2)}$.

If the two particles have spins, their spatial state is either antisymmetric or symmetric depending on the symmetry of total spin state, i.e.,

$$\psi_\pm(X_1, X_2) = \frac{1}{\sqrt{2}}[\varphi(X_1)\phi(X_2) \pm \phi(X_1)\varphi(X_2)].$$

The averaged phase space interaction is $\langle U\rangle_\pm = \langle\psi_\pm(X_1, X_2)|U(\hat{R})|\psi_\pm(X_1, X_2)\rangle \equiv U \pm J$ with the direct interaction U and the exchange interaction J defined by

$$\begin{cases} U \equiv \langle\varphi(X_1)\phi(X_2)|U(\hat{R})|\varphi(X_1)\phi(X_2)\rangle \\ J \equiv \langle\varphi(X_1)\phi(X_2)|U(\hat{R})|\phi(X_1)\varphi(X_2)\rangle. \end{cases} \tag{4.45}$$

The direct interaction part $U = \frac{1}{2}(\langle U\rangle_+ + \langle U\rangle_-)$ corresponds to the classical interaction while the exchange interaction part $J = \frac{1}{2}(\langle U\rangle_+ - \langle U\rangle_-)$ is a pure quantum effect without classical counterpart. In the $\hat{R}$-representation, they have been calculated in appendix B.1, i.e.,

$$U = \sum_N U(R_N)I_N, \quad J = \sum_N(-1)^N U(R_N)I_N \tag{4.46}$$

with the overlap integral

$$I_N = \int dC |\langle \varphi(C + \frac{1}{2}\Delta X)\phi(C - \frac{1}{2}\Delta X)|\Phi_N(\Delta X)\rangle|^2. \tag{4.47}$$

Equations (4.44), (4.46) and (4.47) construct basic framework to calculate quantum phase space interactions.

The expression of phase space interaction (4.44) is based on the unitary transformation (4.41), which means the unperturbed motion of a wave packet is close to an oscillating coherent state. If the unperturbed motion cannot be approximated by the unitary transformation (4.41) or it is difficult to calculate phase space interaction using equations (4.46), (4.50) and (4.51), we can calculate the phase space interaction from the lowest-order Magnus expansion discussed in section 2.3.2. One should always keep in mind that we are investigating the effective stroboscopic dynamics and the two atoms indeed collide with each other during every stroboscopic time step. The phase space interaction is in fact the time-averaged real space interaction over a periodically changing two-body quantum state $\Phi(x_i, x_j, t)$

$$U_{ij} = \frac{\omega}{2\pi}\int_0^{\frac{2\pi}{\omega}} \langle \Phi(x_i, x_j, t)|\, V(x_i - x_j)|\, \Phi(x_i, x_j, t)\rangle dt. \tag{4.48}$$

The above expression is the quantum version of classical phase space interaction given by equation (4.8). This is the method adopted by Sacha *et al* in their study of bouncing atom on an oscillating mirror, see section 3.1.1 and [6–13].

Coherent-state representation It is possible to get an explicit analytical expression of the overlap integral I_N for two displaced squeezed coherent states $\varphi(X)$ and $\phi(X)$ described by

$$\begin{cases} \varphi(X) = \psi_0\left(X - \dfrac{R}{2}\right) = \left(\dfrac{\beta}{\sqrt{\pi}}\right)^{\frac{1}{2}} e^{-\frac{1}{2}\beta^2\left(X - \frac{R}{2}\right)^2} \\ \phi(X) = \psi_0\left(X + \dfrac{R}{2}\right) = \left(\dfrac{\beta}{\sqrt{\pi}}\right)^{\frac{1}{2}} e^{-\frac{1}{2}\beta^2\left(X + \frac{R}{2}\right)^2} \end{cases} \tag{4.49}$$

where R is the distance between the centres of two coherent states in phase space. We refer the readers to the detailed calculation in appendix B.2. The overlap integral I_N is

$$I_N = \frac{\left(\frac{1}{2}\tanh r\right)^N}{N!\cosh r}\left| H_N\left[\frac{\gamma^* e^{i\theta}\sinh r - \gamma\cosh r}{\sqrt{e^{i(\theta+\pi)}\sinh(2r)}}\right]\right|^2 e^{-|\gamma|^2 + \frac{1}{2}(\gamma^{*2}e^{i\theta} + \gamma^2 e^{-i\theta})\tanh r}. \tag{4.50}$$

Here, $\gamma = -R/2\sqrt{\lambda}$ is the displacement parameter, $r = \operatorname{arccosh}(v)$ and $e^{i\theta} = -u/\sinh r$ are the squeezing parameters with v and u given by

$$v = \frac{1}{2}\left(\sqrt{\lambda}\,\beta + \frac{1}{\sqrt{\lambda}\,\beta}\right),\quad u = \frac{1}{2}\left(\sqrt{\lambda}\,\beta - \frac{1}{\sqrt{\lambda}\,\beta}\right). \tag{4.51}$$

For the standard coherent state with $\beta = 1/\sqrt{\lambda}$, the overlap integral of two standard coherent states I_N^s can be simplified into

$$I_N^s = \frac{|\gamma|^{2N}}{N!}e^{-|\gamma|^2} = \frac{1}{N!}\left(\frac{R^2}{4\lambda}\right)^N \exp\left(-\frac{R^2}{4\lambda}\right). \tag{4.52}$$

Note that the distance between two coherent states R is different from the quantised phase space distance R_N. For two overlapped coherent states, their distance R is zero but R_N is always positive as shown by equation (4.39). Given the parameters λ, R and β of two coherent states, one can calculate analytically the direct and exchange interactions U and J from equations (4.46), (4.50) and (4.51).

4.2.2.2 *Collision-induced spin–spin interaction*

Due to the indistinguishability in quantum dynamics, we should reconstruct the symmetric and antisymmetric spatial wave functions of two-body quantum state

$$\Phi^{\pm}(t) = \frac{1}{\sqrt{2N^{\pm}}}\left[\phi(x_i, t)\varphi(x_j, t) \pm \varphi(x_i, t)\phi(x_j, t)\right] \tag{4.53}$$

with normalised factors $N^{\pm}$. Meanwhile, the internal spin states of two-body system can be expressed by the basis of the triplet states $|\downarrow\downarrow\rangle$, $|\uparrow\uparrow\rangle$, $|+\rangle \equiv \frac{1}{\sqrt{2}}(|\uparrow\downarrow\rangle + |\downarrow\uparrow\rangle)$ and the singlet state $|-\rangle \equiv \frac{1}{\sqrt{2}}(|\uparrow\downarrow\rangle - |\downarrow\uparrow\rangle)$. The total states of two particles are the product of the spatial states and the internal spin states. For instance, if the two particles are bosons, the total state needs to be symmetric, resulting in the symmetric basis describing the total states of two ions, i.e., $\{|\Phi^+\rangle\otimes|\downarrow\downarrow\rangle, |\Phi^+\rangle\otimes|\uparrow\uparrow\rangle, |\Phi^+\rangle\otimes|+\rangle, |\Phi^-\rangle\otimes|-\rangle\}$.

In the left part of figure 4.5, we show the internal level structure of two particles in the free space without consideration of their interaction. In this case, the two internal states $|+\rangle$ and $|-\rangle$ are degenerate. When the two particles are colliding each other in the trap, their interaction modifies the internal energy levels of two particles as illustrated by the right part in the same figure. When the kinetic energy of two particles is much larger than the interaction potential, we calculate the level shifts of states $|+\rangle$ and $|-\rangle$ from equation (4.48)

$$\begin{cases} U_{ij}^{+} = \dfrac{\omega}{2\pi}\displaystyle\int_0^{\frac{2\pi}{\omega}} \langle\Phi^+(t)\,|\,V(x_i - x_j)\,|\,\Phi^+(t)\rangle dt \\ U_{ij}^{-} = \dfrac{\omega}{2\pi}\displaystyle\int_0^{\frac{2\pi}{\omega}} \langle\Phi^-(t)\,|\,V(x_i - x_j)\,|\,\Phi^-(t)\rangle dt. \end{cases} \tag{4.54}$$

As shown in the right part of figure 4.5, all the levels of triplet states are shifted by $U + J$ while the level of singlet state is shifted by $U - J$. As a result, the degeneracy of

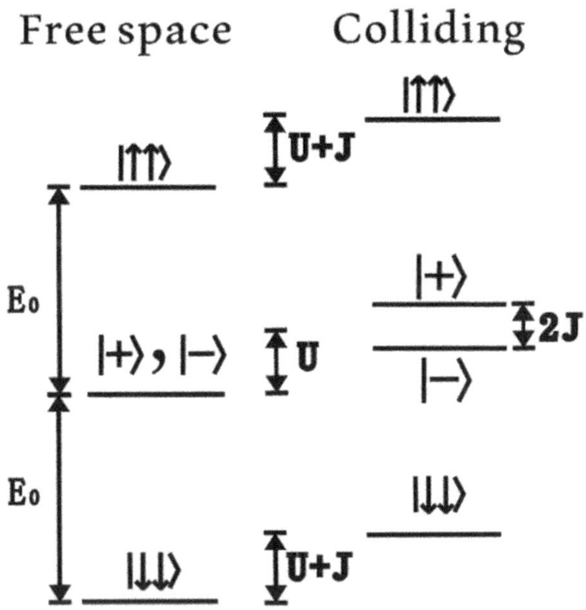

Figure 4.5. Collision-induced spin–spin interaction. The collision of two particles induces a common level shift U and a level splitting $2J$ between the triplet state $|+\rangle$ and singlet state $|-\rangle$. E_0 is the hyperfine level splitting of single particle.

states $|+\rangle$ and $|-\rangle$ is lifted by a level splitting $2J$ with $J \equiv \frac{1}{2}(U_{ij}^{+} - U_{ij}^{-})$. The exchange interaction J here can be viewed as the *collision-induced spin–spin interaction*, which is similar to the effective spin–spin interaction in the Heisenberg model rooting from the static exchange interaction of two neighbouring electrons. The collision-induced direct interaction with $U \equiv \frac{1}{2}(U_{ij}^{+} + U_{ij}^{-})$ is the classical counterpart of the time-averaged direct interaction.

We can use the internal states of two particles as qubit states and design a two-qubit quantum gate for quantum computation. From the relationships $|\downarrow\uparrow\rangle = \frac{1}{\sqrt{2}}(|+\rangle - |-\rangle)$ and $|\uparrow\downarrow\rangle = \frac{1}{\sqrt{2}}(|+\rangle + |-\rangle)$, the time evolutions of initial states $|\downarrow\uparrow\rangle$ and $|\uparrow\downarrow\rangle$ are given by

$$\frac{1}{\sqrt{2}}\left(e^{i\frac{E_0+U+J}{\hbar}t}|+\rangle - e^{i\frac{E_0+U-J}{\hbar}t}|-\rangle\right) \quad \text{and} \quad \frac{1}{\sqrt{2}}\left(e^{i\frac{E_0+U+J}{\hbar}t}|+\rangle + e^{i\frac{E_0+U-J}{\hbar}t}|-\rangle\right)$$

respectively; while the time evolutions of initial states $|\downarrow\downarrow\rangle$ and $|\uparrow\uparrow\rangle$ are simply given by $e^{i\frac{U+J}{\hbar}t}|\downarrow\downarrow\rangle$ and $e^{i\frac{2E_0+U+J}{\hbar}t}|\uparrow\uparrow\rangle$ respectively. By choosing the evolution time $t_g \equiv \frac{3\pi\hbar}{4J}$, the four two-qubit initial states $|\downarrow\downarrow\rangle$, $|\downarrow\uparrow\rangle$, $|\uparrow\downarrow\rangle$ and $|\uparrow\uparrow\rangle$ arrive at

$$
\begin{cases}
|\downarrow\downarrow\rangle & \mapsto & e^{i\frac{(U+J)}{\hbar}t_g}\,|\downarrow\downarrow\rangle \\
|\downarrow\uparrow\rangle & \mapsto & e^{i\frac{(E_0+U+J)}{\hbar}t_g}\frac{1}{2}[(1+i)|\downarrow\uparrow\rangle + (1-i)|\uparrow\downarrow\rangle] \\
|\uparrow\downarrow\rangle & \mapsto & e^{i\frac{(E_0+U+J)}{\hbar}t_g}\frac{1}{2}[(1-i)|\downarrow\uparrow\rangle + (1+i)|\uparrow\downarrow\rangle] \\
|\uparrow\uparrow\rangle & \mapsto & e^{i\frac{(2E_0+U+J)}{\hbar}t_g}\,|\uparrow\uparrow\rangle.
\end{cases}
\tag{4.55}
$$

In the basis of $\{|\downarrow\downarrow\rangle, |\downarrow\uparrow\rangle, |\uparrow\downarrow\rangle, |\uparrow\uparrow\rangle\}$, the above unitary time evolution defines a two-qubit quantum gate

$$
\mathcal{U}(t_g) \equiv e^{i\frac{(E_0+U+J)}{\hbar}t_g}
\begin{pmatrix}
e^{-i\frac{E_0}{\hbar}t_g} & 0 & 0 & 0 \\
0 & \frac{1}{2}(1+i) & \frac{1}{2}(1-i) & 0 \\
0 & \frac{1}{2}(1-i) & \frac{1}{2}(1+i) & 0 \\
0 & 0 & 0 & e^{i\frac{E_0}{\hbar}t_g}
\end{pmatrix}.
\tag{4.56}
$$

The two-qubit gate $\mathcal{U}(t_g) = \sqrt{SWAP}(R_\theta \otimes R_\theta)$ can be decomposed into the standard two-qubit $\sqrt{SWAP}$ gate [14, 15] and two single-qubit phase shift gates, where the single-qubit phase gate is defined by $R_\theta \equiv \mathrm{diag}(1, \theta)$ with phase parameter $\theta = t_g(2E_0 + U + J)/2\hbar$. As the two-qubit $\sqrt{SWAP}$ gate is universal [14, 15], meaning any quantum algorithm can be realised by the universal two-qubit gate plus single-qubit gates, hence the two-qubit $U(t_g)$ gate introduced here is also universal.

4.2.3 Example interactions

We now calculate analytical expressions of $U(R)$ and $J(R)$ for three concrete example interactions, i.e., the contact interaction, Coulomb interaction and hard-sphere interactions.

4.2.3.1 Contact interaction

In the experiments of ultracold atoms, the collision of cold atoms is dominated by the s-wave scattering process. The two-body interactions of ultracold gases in 3D can be modelled by a pseudopotential in the form of contact function [2] $V_{3D}(\mathbf{r}) = \frac{4\pi\hbar^2 a}{2m}\delta(\mathbf{r})$, where a is the s-wave scattering length. In the case of quasi-1D dimension where the transverse trapping frequency $\omega_\perp$ is much larger than the longitudinal trapping frequency ω_x, the effective pseudopotential is described by an interaction of the form $V_{1D}(x) = \varepsilon\delta(x)$, where $\varepsilon = \frac{2\hbar\omega_\perp a}{1 - Aa/l_\perp}$ is the interaction strength with the constant $A \approx 1.036$ [2]. From $V_q = (2\pi)^{-1}\int_{-\infty}^{+\infty} dx\varepsilon\delta(x)e^{-iqx} = (2\pi)^{-1}\varepsilon$ and equation (4.44), the phase space interaction potential can be calculated [4]

$$U(R_N) = \varepsilon \frac{1 + (-1)^N}{N\pi\sqrt{2\lambda}} \frac{\Gamma\left(\frac{N+1}{2}\right)}{\Gamma\left(\frac{N}{2}\right)}. \tag{4.57}$$

Here, $N = 0, 1, 2, 3 \cdots$ and $\Gamma(\bullet)$ is the gamma function. We find that $U(R_N)$ is zero for odd integer N due to the fact that the wave function of two atoms' relative motion $\Phi_N(\Delta X)$ is antisymmetric for odd N, which means the probability amplitude is zero when the two atoms contact each other. As a result, the total averaged value of point-like contact interaction over a antisymmetric function is zero. The direct phase space interaction $U(R)$ and the exchange interaction $J(R)$ of two standard coherent states can be directly calculated from equations (4.46), (4.52) and (4.57)

$$\begin{cases} U(R) &= \dfrac{\varepsilon}{\sqrt{2\pi\lambda}} \exp\left(-\dfrac{R^2}{4\lambda}\right) I_0\left(\dfrac{R^2}{4\lambda}\right) \\ J(R) &= U(R). \end{cases} \tag{4.58}$$

Here, R is the distance in phase space between the centres of two coherent states and $I_0(\bullet)$ is the zeroth order modified Bessel function of the first kind. Using the asymptotic behaviour of the special function $I_0(z) \sim e^z/\sqrt{2\pi z}$ for $z \gg 1$, we have

$$U(R) \sim \frac{\varepsilon}{\pi} \frac{1}{R}, \quad \text{for } R \gg 2\sqrt{\lambda}, \tag{4.59}$$

which is consistent with the result in classical limit equation (4.12). In figure 4.6(b), we plot $U(R)$ as a function of R and its long-range asymptotic behaviour. Note that equation (4.58) also shows the exchange interaction is equal to the direct phase space interaction.

4.2.3.2 Coulomb interaction

In the example of cold atoms, the equality of direct and exchange interactions $J(R) = U(R)$ comes from the δ function modelling the contact interaction and the

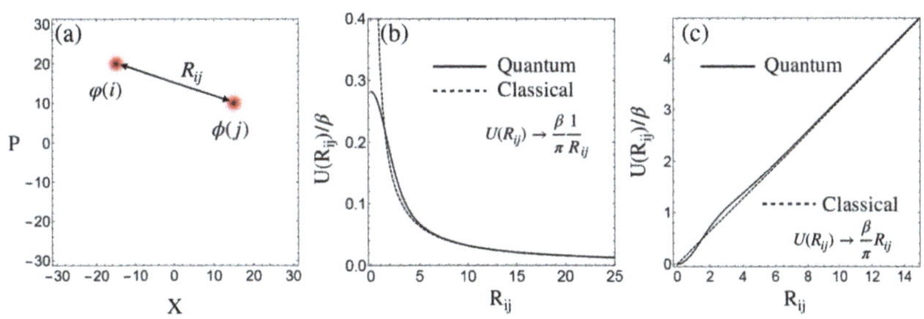

Figure 4.6. Quantum phase space interactions. (a) Two coherent states $\varphi(i)$ and $\phi(j)$ in phase space with distance R_{ij} of their centres. (b) Phase space interaction potential $U(R_{ij})$ for the contact interaction. (c) $U(R_{ij})$ for hard-sphere interaction. In both figures (b) and (c), the black solid curve is the quantum interaction ($\lambda = 2$) and the blue dashed line is the classical interaction ($\lambda \to 0$).

fact that the spatial antisymmetric state of two atoms has zero probability to touch each other. If the interaction potential between particles is different from the contact interaction model, it is possible to tune $U(R)$ and $J(R)$ independently in the experiments. To illustrate this, we replace the two neutral atoms by two charged ions, whose interaction is dominated by the strong Coulomb interaction $V(r) = \frac{Q^2}{4\pi\epsilon_0}\frac{1}{r}$, where Q is the charge of single ion and r is the distance of two ions. Assuming the strong confinement condition $\omega_\perp \gg 1$, the two ions are confined in a quasi-1D dimension and the radial motions of two ions are frozen in the ground states. The average Coulomb interaction of two ions by the ground states in the y–z plane is described by an effective Coulomb interaction in the axial x direction [16]

$$V(r_x) = \frac{Q^2}{4\pi\epsilon_0 x_0}\sqrt{\frac{\pi\omega_\perp}{2}}\exp\left(\frac{\omega_\perp}{2}\frac{r_x^2}{x_0^2}\right)\operatorname{erfc}\left(\sqrt{\frac{\omega_\perp}{2}}\frac{r_x}{x_0}\right), \tag{4.60}$$

where $x_0 = \sqrt{\hbar/m\omega_x}$, $r_x \equiv |x_1 - x_2|$ and erfc[•] is the complementary error function. In the limit of $r_x \gg x_0\sqrt{2/\omega_\perp}$, we have the asymptotic behaviour $V(r_x) \to \frac{Q^2}{4\pi\epsilon_0 r_x}$ recovering Coulomb's law in one dimension. In the limit of zero distance $r_x = 0$, the effective interaction (4.60) has a finite potential barrier $V(0) = \frac{Q^2}{4\pi\epsilon_0 x_0}\sqrt{\frac{\pi\omega_\perp}{2}}$. When the kinetic energy $E_k = \frac{1}{4}m\omega_x^2 R^2$ of two ions described by two coherent states with phase space distance R is much larger than the effective Coulomb interaction potential barrier, i.e., $E_k \gg V(0)$, the collision-induced exchange interaction J and the direct interaction U in the long-distance limit $R \gg x_0\sqrt{\omega_\perp/2}$ was calculated using equation (4.48) [17],

$$J(R) \sim \frac{Q^2}{2\pi^2\epsilon_0}\frac{\hbar\omega_{y,z}}{m\omega_x^2}\frac{1}{R^3} \quad \text{and} \quad U(R) \sim \frac{\tilde{Q}^2}{4\pi\epsilon_0}\frac{1}{R}, \tag{4.61}$$

where $\tilde{Q} \equiv Q\sqrt{2\pi^{-1}\ln(2^{3/2}e^{\gamma/2}\omega_\perp^{1/2}R/x_0)}$ is the renormalised charge with $\gamma = 0.577\,2\cdots$ the Euler–Mascheroni constant. The direct interaction $U(R)$ and exchange interaction $J(R)$ as functions of phase space distance R follow different power laws and can be tuned independently by the trapping frequencies $\omega_{x,y,z}$. Note that J is proportional to $\hbar$ reflecting the quantum feature of exchange interaction, while U is a classical quantity, which keeps the form of Coulomb interaction with a renormalised charge.

4.2.3.3 Hard-sphere interaction

In the pure-1D regime, where the characteristic length of transverse trapping $l_\perp$ is much smaller than the cold atom's size, the contact interaction is no longer valid for the description of interaction between cold atoms [18]. In this situation, the atom can be viewed as a hard-sphere particle with a radius σ, which means the interaction potential between the two atoms is infinite when their distance is smaller than 2σ and zero when the distance is larger than 2σ. Our theory of phase space interaction can be applied to the small hard-sphere limit $\sigma \ll \sqrt{\lambda}$. Since the two atoms cannot contact each other due to the hard-sphere interaction, the eigenstates of phase space distance operator $\hat{R}$ have to be zero at zero distance, which means that only the odd eigenstates $\Phi_{2m+1}(\Delta X)$ with $m \in \mathbb{N}$ satisfy this condition. The even eigenstate should be reconstructed as $\Phi_{2m}(\Delta X) = \mathrm{sgn}(\Delta X)\Phi_{2m+1}(\Delta X)$ with $\mathrm{sgn}(\bullet)$ the sign function. The eigenstates $\Phi_{2m}(\Delta X)$ and $\Phi_{2m+1}(\Delta X)$ are degenerate with the same eigenvalue $R_{2m} = R_{2m+1} = 2\sqrt{\lambda(2m + 3/2)}$. In appendix B.3, we calculate the phase space interaction potential of the hard-sphere interaction potential for odd integers $N = 2m + 1$

$$U(R_N) = \frac{\sigma\sqrt{2\lambda/\pi}}{2^{-N}N!}\sum_{k,l=0}^{\left[\frac{N}{2}\right]\left[\frac{N}{2}\right]}\frac{(-1)^{k+l}(N!)^2(N-k-l)!}{2^{2k+2l}k!l!(N-2k)!(N-2l)!} \approx \frac{2\sigma}{\pi}R_N. \tag{4.62}$$

Here, $[\frac{N}{2}]$ means the closest integer number less than $\frac{N}{2}$. For even integers $N = 2m$, we have $U(R_{2m}) = U(R_{2m+1})$. Here, we find that $U(R_N)$ can be approximated very well by the linear relationship $U(R_N) \approx 2\sigma\pi^{-1}R_N$.

The direct phase space interaction $U(R)$ and the exchange interaction $J(R)$ of two coherent states can be calculated from equations (4.46), (4.62) [4]

$$\begin{cases} U(R) &= 2\sum_{m=0} U(R_{2m+1})I_{2m+1} \\ J(R) &= 0. \end{cases} \tag{4.63}$$

Here, R is phase space distance between the centres of two coherent states and the overlap integral I_{2m+1} is given by equation (4.52). The zero exchange interaction comes from the degeneracy of the symmetric and antisymmetric states. Using the linear approximation $U(R_N) \approx 2\sigma\pi^{-1}R_N$ and equation (4.63), we have

$$U(R) \approx \frac{8\sigma\sqrt{\lambda}}{\pi}\sum_{m=0}^{\infty}\frac{\sqrt{2m+3/2}}{(2m+1)!}\exp\left(-\frac{R^2}{4\lambda}\right)\left(\frac{R^2}{4\lambda}\right)^{2m+1}. \tag{4.64}$$

In the long-distance limit, we have the asymptotic expression of equation (4.64), i.e.,

$$U(R) \to \frac{2\sigma}{\pi}R, \quad \text{for } R \gg 2\sqrt{\lambda}. \tag{4.65}$$

This is consistent again with the classical result equation (4.16) of power-law interaction for $n \to \infty$. In figure 4.6(c), we plot the direct phase space interaction potential $U(R)$ as a function of phase space distance R. We see that the linear relationship equation (4.65) (dashed line) is a very good approximation of equation (4.63) (solid curve). It is interesting to find that the linear phase space interaction potential equation (4.65) mimics the interaction potential between quarks in QCD [19, 20]. This surprising behaviour of hard-sphere atoms can be understood in a simple picture. Since the two atoms have a tiny hard-sphere radius σ, they prefer to oscillate in a synchronised way (in-phase oscillation). If the atoms' oscillations are out of phase, due to the finite phase space distance R, they are more likely to collide with each other during the oscillation. The collision effect becomes stronger as the phase space distance R is larger, resulting in a confinement potential in the end.

4.2.4 Strongly correlated models in phase space

We have established the relationship between the effective interaction in phase space and the original interaction in real space, both in classical limit and quantum regime. A key finding is that a long-range phase space interaction can be created (e.g., Coulomb-like form) by a short-range real space interaction (e.g., point-like form). Based on the idea of synthesising phase space lattices via driving field and the long-range phase space interaction, we can go beyond single-particle physics and study condensed matter physics in phase space. In this section, we will discuss how strongly correlated models, e.g., Hubbard model, many-body Anderson model and Heisenberg model, are simulated in phase space.

4.2.4.1 Hubbard model

In section 3.2.2, we introduced the tight-binding model (3.22) based on the framework of single-particle band theory. With the theory of phase space interaction, we can extend the tight-binding model (3.22) for free particles to the many-body Hamiltonian with interactions. For spinless Bosonic particles, the many-body Hamiltonian in phase space is

$$\begin{aligned} H_{\mathrm{BH}} \approx E_l \sum_i \hat{n}_{i,l} + \frac{1}{2}\left(\sum_{\langle i,j\rangle} J_l \hat{a}^{\dagger}_{i,\,l} \hat{a}_{j,l} + h.\,c.\right) + \sum_i \frac{1}{2} U_{ii,l} \hat{n}_{i,l}(\hat{n}_{i,l} - 1) \\ + \frac{1}{2}\sum_{i \neq j} U_{ij,l} \hat{n}_{i,l} \hat{n}_{j,l}. \end{aligned} \tag{4.66}$$

Here, $\hat{a}^{\dagger}_{i,\,l}$ ($\hat{a}_{i,l}$) is the creating (annihilating) operator of the ith Wannier state in the lth band and $\hat{n}_{i,l} = \hat{a}^{\dagger}_{i,\,l}\hat{a}_{i,l}$ is the number operator occupying the Fock space $|\dots, n_{i,l}, \dots\rangle$ in the Wannier representation. The parameter J_l is the tunnelling rate between nearest lattice sites, which is a complex number in general, and is given by equation (3.24). We have neglected the tunnelling rates between lattices beyond nearest-neighbours since the tunnelling amplitude is usually exponentially suppressed with increasing phase space distance. As a contrast, due to the long-range feature of phase space interactions, we keep all the interaction terms $U_{ij,l}$ in the Hamiltonian (4.66). For the lowest tight-binding band, the Wannier states can be approximated by coherent states [2], one can directly use the analytical expressions equations (4.58), (4.61) and (4.64) for different interaction potentials. For higher bands, the phase space interaction $U_{ij,l}$ may not have a compact form. We have to use equation (4.48) that is expressed in the laboratory frame

$$U_{ij,l} = \int_0^T \mathrm{d}t \langle \phi_{i,l}(x_m, t)\phi_{j,l}(x_n, t)| V(x_m - x_n) |\phi_{i,l}(x_m, t)\phi_{j,l}(x_n, t)\rangle, \tag{4.67}$$

where $\phi_{i,l}(x_m, t)$ represents the lth Wannier wave function on the phase space lattice site i for the mth particle in the laboratory frame. For the contact interaction $V(x_m - x_n) = g_0\delta(x_m - x_n)$, we have $U_{ij,l} = g_0 \int_0^T \mathrm{d}t \int \mathrm{d}x|\phi_{i,l}(x, t)|^2|\phi_{j,l}(x, t)|^2$ [6–13]. Model (4.66) is valid when the interaction $U_{ij,l}$ is weak compared to the quasienergy gaps.

By neglecting the off-site interaction terms $U_{i\neq j,l}$, Hamiltonian (4.66) is the paradigm Bose–Hubbard model for the study of strongly correlated phenomena in condensed matter physics. In [6], Sacha *et al* discussed Mott transition using the model of bouncing atoms on an oscillating mirror. In the limit of free interaction, the ground state of many-body Hamiltonian (4.66) is a superfluid state with long-time phase coherence. However, for sufficiently strong repulsive contact interaction ($g_0 > 0$) such that $U_{ii,l} \ll N|J_l|/n$ (neglecting off-site interaction terms), the ground state of (4.66) is a single Fock state with well-defined numbers of atoms occupying each Wannier state, i.e., $|N/n, N/n, \ldots, N/n\rangle$ with N denoting the total number of particles. As a result, a gap opens between the ground state level and excited levels, and consequently a Mott insulator phase is achieved. Sacha *et al* also investigated the model (4.66) in two dimensions by extending atoms bouncing between two orthogonal harmonically oscillating mirrors [11, 21].

It is also direct to extend the Bose–Hubbard Hamiltonian (4.66) to the Fermi–Hubbard model

$$H_{\mathrm{FH}} \approx \sum_{i,\sigma} E_i\hat{n}_{i,\sigma} + \frac{1}{2}\left(\sum_{\langle i,j\rangle} J_l\hat{c}^{\dagger}_{i,\sigma}\hat{c}_{j,\sigma} + h.\ c.\right) + \sum_i \frac{1}{2}U_{ii}\hat{n}_{i,\sigma}\hat{n}_{i,\bar{\sigma}}. \tag{4.68}$$

Here, we have restricted to the case of single band and included the Pauli exclusion principle in the interaction terms, i.e., one lattice site can be occupied by at most two particles with opposite spins (σ and $\bar{\sigma}$).

By including off-site interactions in (4.66), more intriguing condensed matter phenomena, e.g., the topological Haldane phase, are possible to be explored [8].

4.2.4.2 Many-body Anderson model

In section 3.2.4, we introduced the Anderson model and Anderson localisation for a single particle. It is quite natural to combine interactions between particles with temporal disorder. This extends the analysis of Anderson localisation to interacting particle systems, i.e., the many-body localisation (MBL).

In [12], Sacha *et al* proposed the following Bose–Hubbard Hamiltonian with on-site disorder

$$H_{\mathrm{eff}} \approx -\frac{J}{2}\sum_{j=1}^{n}(\hat{a}^{\dagger}_{j+1}\hat{a}_j + \mathrm{c.\ c.}) + \frac{1}{2}\sum_{i,j}^{n} U_{ij}\hat{a}^{\dagger}_i\hat{a}_i\hat{a}^{\dagger}_j\hat{a}_j + \sum_{j=1}^{n}\epsilon_j\hat{a}^{\dagger}_j\hat{a}_j, \tag{4.69}$$

where the interaction U_{ij} and on-site disorder ϵ_j are given by equation (4.67) and equation (3.38) respectively. The Hamiltonian (4.69) is valid provided the interaction energy NU_{ij} (N is a total number of bosons) and the disorder ϵ_j are much smaller than the energy gap. Although the disorder breaks the translational invariance of the phase space lattice, the system still possesses Floquet eigenstates with longer period $2\pi n/\omega_d$ equal to the period of disorder, i.e., $H'(t + nT_d) = H'(t)$. The MBL in the disordered Hubbard model has been observed in the experiment [22]. It is believed that MBL is constrained to systems with short-range interactions [23] but the long-range interaction alone does not exclude the MBL [24]. The existence of MBL with long-range interactions is an important but unexplored problem. In the experiments with cold atoms, the interaction range is short since the atoms are neutral in general. The many-body Hamiltonian (4.69) from subresonantly driven protocol may present a new approach to solve this problem, as the long-range interaction U_{ij} can be generated from contact interaction directly.

By modulating the strength of contact interaction with a proper temporal disorder, the particles can have an effective disordered interaction potential in phase space. In such a way, a two-particle molecule bound by Anderson localisation is created from the periodic driving, which is called the *Anderson molecule* [13, 25].

4.2.4.3 Heisenberg model

In section 4.2.2.2, we discussed the collision-induced spin–spin interaction, which is the exchange interaction energy. For the contact interaction, it is found that the exchange interaction is equal to the direct interaction, i.e., $J_{ij} = J(R_{ij}) = U(R_{ij})$. For Coulomb interaction, the exchange interaction is $J(R_{ij}) \propto \hbar/R_{ij}^3$ which decays in the form of the inverse-cube law. For the hard-sphere interaction, the particles cannot cross each other, and thus the exchange interaction is zero $J_{ij} = 0$.

If the particles are tightly bound at their equilibrium points by the phase space lattice, the direct phase space interaction $U(R_{ij})$ does not play a role in the dynamics. The dynamics of the spins is determined by the exchange interaction and is described by the Heisenberg model with isotropic spin–spin interaction, i.e.,

$$H_{\text{ex}} = -\frac{1}{4}\sum_{i\neq j} J(R_{ij})\hat{S}_i \cdot \hat{S}_j + \text{constant}. \tag{4.70}$$

In figure 4.7, we show the Heisenberg models in the $\mathbb{Z}_n$ circular lattice and square lattice in phase space. The operator $\hat{S}_i \equiv \sum_m \hat{s}_{im}$ is the total spin operator on the ith site of phase space lattice since more than one spins can be confined in one lattice site. For the $\mathbb{Z}_n$ circular lattice, the origin point (0, 0) in phase space is also a stable point, which can hold a central spin interacting with the spins on the $\mathbb{Z}_n$ lattice. Therefore, the corresponding Heisenberg model up to a constant is

$$H_{\text{ex}} = -\frac{1}{4}\sum_{i\neq j}^{n} J(R_{ij})\hat{S}_i \cdot \hat{S}_j - \frac{1}{4}\sum_{i=1}^{n} J(R_{i0})\hat{S}_0 \cdot \hat{S}_i, \tag{4.71}$$

where $\hat{S}_{i(j)}$ with $i(j) = 1, 2, \ldots, n$ represents the spins on the $\mathbb{Z}_n$ lattice sites and $\hat{S}_0$ is the central spin. This model is closely related to the central spin models [26, 27].

The famous Mermin–Wagner theorem claims that a 1D or 2D isotropic Heisenberg model with finite-range exchange interaction can be neither

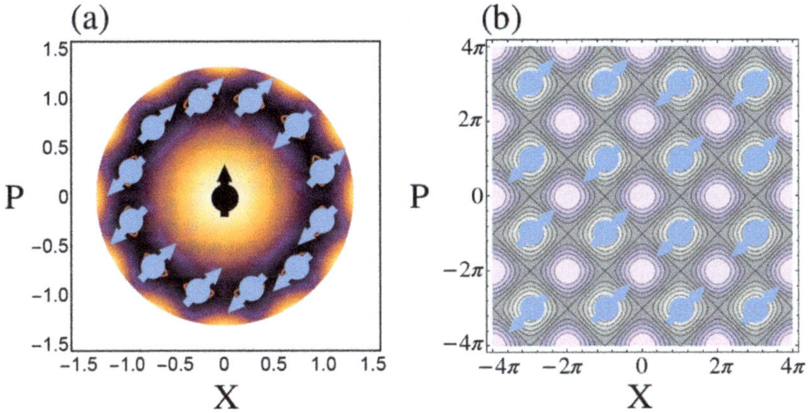

Figure 4.7. Heisenberg models. (a) 1D Heisenberg model of periodically arranged spins (blue) on the $\mathbb{Z}_n$ phase space lattice sites interacting with a central spin (black) in the origin. (b) 2D Heisenberg model of spins occupying one sublattice of the square phase space lattice. Reprinted from [39].

ferromagnetic nor antiferromagnetic [28] at any nonzero temperature, which clearly excludes a variety of types of long-range ordering in low dimensions. Here in our model, the Floquet exchange interaction from point-like contact interaction has a Coulomb-like long-range behaviour. Therefore, the phase space lattices provide a possible platform to test the Mermin–Wagner theorem and search for other new phenomena with long-range interactions, such as causality and quantum criticality [29–35], nonlocal order [36–38], etc.

References

[1] Guo L, Liu M and Marthaler M 2016 Effective long-distance interaction from short-distance interaction in a periodically driven one-dimensional classical system *Phys. Rev.* A **93** 053616

[2] Bloch I, Dalibard J and Zwerger W 2008 Many-body physics with ultracold gases *Rev. Mod. Phys.* **80** 885–964

[3] Guo L and Marthaler M 2016 Synthesizing lattice structures in phase space *New J. Phys.* **18** 023006

[4] Liang P, Marthaler M and Guo L 2018 Floquet many-body engineering: topology and many-body physics in phase space lattices *New J. Phys.* **20** 023043

[5] Wünsche A 1991 Displaced Fock states and their connection to quasiprobabilities *Quantum Opt.* **3** 359–83

[6] Sacha K 2015 Anderson localization and Mott insulator phase in the time domain *Sci. Rep.* **5** 10787

[7] Sacha K 2015 Modeling spontaneous breaking of time-translation symmetry *Phys. Rev.* A **91** 033617

[8] Giergiel K, Dauphin A, Lewenstein M, Zakrzewski J and Sacha K 2019 Topological time crystals *New J. Phys.* **21** 052003

[9] Kosior A and Sacha K 2018 Dynamical quantum phase transitions in discrete time crystals *Phys. Rev.* A **97** 053621

[10] Giergiel K, Kosior A, Hannaford P and Sacha K 2018 Time crystals: analysis of experimental conditions *Phys. Rev.* A **98** 013613

[11] Giergiel K, Kuroś A and Sacha K 2019 Discrete time quasicrystals *Phys. Rev.* **B 99** 220303

[12] Mierzejewski M, Giergiel K and Sacha K 2017 Many-body localization caused by temporal disorder *Phys. Rev.* B **96** 140201

[13] Giergiel K, Miroszewski A and Sacha K 2018 Time crystal platform: from quasicrystal structures in time to systems with exotic interactions *Phys. Rev. Lett.* **120** 140401

[14] Loss D and DiVincenzo D P 1998 Quantum computation with quantum dots *Phys. Rev.* A **57** 120–6

[15] Schuch N and Siewert J 2003 Natural two-qubit gate for quantum computation using the XY interaction *Phys. Rev.* A **67** 032301

[16] Yin J and Javanainen J 1995 Quantum motion of two trapped ions in one dimension *Phys. Rev.* A **51** 3959–66

[17] Liang P and Guo L 2018 Scalable ion trap architecture for universal quantum computation by collisions arXiv:1801.02593

[18] Astrakharchik G E, Blume D, Giorgini S and Granger B E 2004 Quasi-one-dimensional Bose gases with a large scattering length *Phys. Rev. Lett.* **92** 030402

[19] Trawiński A P, Głazek S D, Brodsky S J, de Téramond G F and Dosch H G 2014 Effective confining potentials for QCD *Phys. Rev.* D **90** 074017

[20] Jido D and Sakashita M 2016 Quark confinement potential examined by excitation energy of the Λ c and Λ b baryons in a quark–diquark model *Prog. Theor. Exp. Phys.* **2016** 083D02

[21] Giergiel K, Miroszewski A and Sacha K 2018 Time crystal platform: From quasicrystal structures in time to systems with exotic interactions *Phys. Rev. Lett.* **120** 140401

[22] Schreiber M, Hodgman S S, Bordia P, Lüschen H P, Fischer M H, Vosk R, Altman E, Schneider U and Bloch I 2015 Observation of many-body localization of interacting fermions in a quasirandom optical lattice *Science* **349** 842–5

[23] Yao N Y, Laumann C R, Gopalakrishnan S, Knap M, Müller M, Demler E A and Lukin M D 2014 Many-body localization in dipolar systems *Phys. Rev. Lett.* **113** 243002

[24] Nandkishore R M and Sondhi S L 2017 Many-body localization with long-range interactions *Phys. Rev.* X **7** 041021

[25] Matus P, Giergiel K and Sacha K 2021 Anderson complexes: Bound states of atoms due to Anderson localization *Phys. Rev.* A **103** 023320

[26] Claeys P W, De Baerdemacker S, Araby O E and Caux J-S 2018 Spin polarization through Floquet resonances in a driven central spin model *Phys. Rev. Lett.* **121** 080401

[27] Ashida Y, Shi T, Schmidt R, Sadeghpour H R, Cirac J I and Demler E 2019 Quantum rydberg central spin model *Phys. Rev. Lett.* **123** 183001

[28] Mermin N D and Wagner H 1966 Absence of ferromagnetism or antiferromagnetism in one- or two-dimensional isotropic Heisenberg models *Phys. Rev. Lett.* **17** 1133–6

[29] Hauke P and Tagliacozzo L 2013 Spread of correlations in long-range interacting quantum systems *Phys. Rev. Lett.* **111** 207202

[30] Richerme P, Gong Z-X, Lee A, Senko C, Smith J, Foss-Feig M, Michalakis S, Gorshkov A V and Monroe C 2014 Non-local propagation of correlations in quantum systems with long-range interactions *Nature* **511** 198–201

[31] Gong Z-X, Foss-Feig M, Michalakis S and Gorshkov A V 2014 Persistence of locality in systems with power-law interactions *Phys. Rev. Lett.* **113** 030602

[32] Métivier D, Bachelard R and Kastner M 2014 Spreading of perturbations in long-range interacting classical lattice models *Phys. Rev. Lett.* **112** 210601

[33] Foss-Feig M, Gong Z-X, Clark C W and Gorshkov A V 2015 Nearly linear light cones in long-range interacting quantum systems *Phys. Rev. Lett.* **114** 157201

[34] Maghrebi M F, Gong Z-X, Foss-Feig M and Gorshkov A V 2016 Causality and quantum criticality in long-range lattice models *Phys. Rev.* **B 93** 125128

[35] Buyskikh A S, Fagotti M, Schachenmayer J, Essler F and Daley A J 2016 Entanglement growth and correlation spreading with variable-range interactions in spin and fermionic tunneling models *Phys. Rev.* A **93** 053620

[36] Dalla Torre E G, Berg E and Altman E 2006 Hidden order in 1d Bose insulators *Phys. Rev. Lett.* **97** 260401

[37] Berg E, Dalla Torre E G, Giamarchi T and Altman E 2008 Rise and fall of hidden string order of lattice bosons *Phys. Rev.* B **77** 245119

[38] Endres M, Cheneau M, Fukuhara T and Weitenberg C 2011 Observation of correlated particle–hole pairs and string order in low-dimensional Mott insulators *Science* **334** 200–3

[39] Guo L and Liang P 2020 Condensed matter physics in time crystals *New J. Phys.* **22** 075003

Chapter 5

Phase space crystal vibrations

In this chapter, we discuss the linearised dynamics of a 2D phase space crystal. The phase space interactions between atoms give rise to collective vibrational band structure with non-trivial Chern numbers and chiral edge states in phase space [1]. Remarkably, these arise despite the presence of time-reversal symmetry, in contrast to all topological transport scenarios in real space. In section 5.1, we introduce the many-body Hamiltonian of honeycomb phase space crystal with long-range Coulomb-type interaction and its linearised lattice wave model. The vibrations of any two atoms are coupled via a complex pairing interaction. The complex phases of paring couplings have a simple geometrical interpretation and cannot be eliminated by a gauge transformation. In section 5.2, we investigate the topological aspects of bulk bands and the robust edge states with different open boundary conditions. Phase space crystals support Chern insulators topological phases analogous to anomalous quantum Hall effects of 2D systems but with preserved time-reversal symmetry. Phase space crystal vibrations offer a new platform for phononic topological transport, with chiral properties arising naturally.

5.1 Lattice dynamics

5.1.1 Honeycomb phase space lattice

We consider cold atoms trapped in a quasi-1D elongated harmonic potential with the axial trapping frequency $\omega_{\rm ax}$ and transverse trapping frequency $\omega_{\rm tr} \gg \omega_{\rm ax}$. In the experiment, one can start from a small Bose–Einstein condensate (BEC) in a magnetic trap [2]. Then, the BEC is loaded into a 2D optical potential along the y and z directions, as shown by figure 5.1(a), by superimposing two orthogonal standing waves on top of the BEC. Each standing wave is formed by two counter-propagating Gaussian laser beams. Supposing the laser light has wavelength $\lambda_L = 2\pi/k_L$, the lattice potential has the form of $V(x, y, z) = V_0(\sin^2 ky + \sin^2 kz)$ with the potential depth V_0 laser intensity. As a result, an array of 1D quantum gases confined to narrow potential tubes is created. For a sufficiently strong potential

doi:10.1088/978-0-7503-3563-8ch5

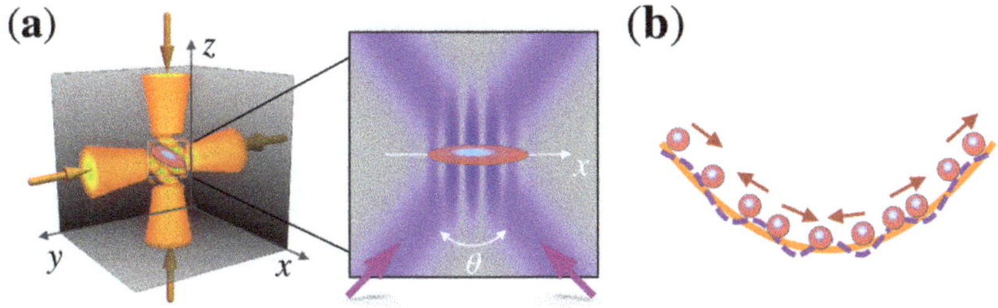

Figure 5.1. Experimental implementation of model Hamiltonian (5.1) with cold atoms. (a) Atomic cloud (red cigar) confined in a quasi-1D tube by two orthogonal polarised laser fields (orange) and tunable stroboscopic lattice by two laser beams (purple) intersecting at angle θ. (b) Sketch of our model: particles (red dots) confined in a static harmonic potential (orange solid curve) and driven by stroboscopic lattices (purple dashed curve). Produced using own data but following methods in [1].

depth (laser intensity), the tunnel coupling and particle exchange between different tubes are exponentially suppressed [3]. The Gaussian profile of the laser beams also leads to axial confinement of the quasi-1D gases. The resulting transverse (in the y–z plane) trapping frequency ω_{tr} and axial trapping frequency ω_{ax} are given by [4] $\omega_{\text{tr}} = \frac{2E_r}{\hbar}\sqrt{\frac{V_0}{E_r}}$, $\omega_{\text{ax}} = \frac{\lambda_L}{\pi w_0}\omega_{\text{tr}}$, where $E_r = \hbar^2 k_L^2/2m$ is the recoil energy of an atom with mass m, and w_0 is the Gaussian beam waist that sets the length of 1D harmonic trap.

We further superimpose two equally polarised laser beams of wavelength λ_D intersecting at an angle θ as shown by the left panel of figure 5.1(a). The result is a standing wave optical dipole potential with a spatial period of [5] $d = \lambda_D/2\sin(\theta/2)$. We can control the stroboscopic lattice constant d, which needs to be much longer than the wavelength of laser lights for our purpose, by tuning the angle θ. The lattice potential can also be tuned on and off with time period $\tau\omega_{\text{ax}}^{-1}$. The single atom dynamics is described by the Hamiltonian [1]

$$H_s(x, p) = \frac{1}{2}(x^2 + p^2) + \sum_{n\in\mathbb{Z}}\sum_q K_q \cos(k_q x - \phi_q)\delta\left(\frac{t}{\tau} - \theta_q - n\right). \tag{5.1}$$

Here, we have rescaled the coordinate, momentum, time and energy by k^{-1}, $m\omega_{\text{ax}}/k$, ω_{ax}^{-1} and $m\omega_{\text{ax}}^2/k^2$, respectively, where m is the mass and k the typical wave vector of the stroboscopic lattices. Hamiltonian (5.1) is the generalised model of kicked harmonic oscillator (KHO) [6–11] including multiple stroboscopic optical lattices labelled by q. At each kicking time instant $t = (n + \theta_q)\tau$ with $n \in \mathbb{Z}$, the kicking strength K_q, lattice wave vector k_q and phase θ_q are tunable parameters. We sketch our model in figure 5.1(b).

For weak near-resonant driving ($|K_q| \ll 1$, $|\tau - 2\pi| \ll 1$), the single-particle dynamics is dominated by fast harmonic oscillations with slow changing quadratures (X, P): $x = P\sin(2\pi t/\tau) + X\cos(2\pi t/\tau)$. Within a rotating-wave approximation (RWA) this leads to a time-independent phase space Hamiltonian $H_s(X, P)$. The extrema of $H_s(X, P)$ represent stable points of the classical dynamics. For an appropriate choice of parameters, they will form a lattice in phase space. In fact, any arbitrary lattice Hamiltonian can be synthesised by choosing proper driving

Table 5.1. Kicking parameters for a honeycomb lattice.

θ_q	K_q	k_q	ϕ_q
0, 1/6, 1/3	Λ	2	0
1/12, 5/12, 9/12	$\Lambda/2$	$4/\sqrt{3}$	π
3/12, 7/12, 11/12	Λ	$2/\sqrt{3}$	π

parameters (section 3.1.2 and appendix A). In order to create a honeycomb lattice in phase space, we need a kicking sequence composed of nine stroboscopic lattices with parameters given by table 5.1. There are three groups of stroboscopic lattices with the ratio of lattice constants

$$d_1\colon d_2\colon d_3 = \frac{1}{2}\colon \frac{\sqrt{3}}{4}\colon \frac{\sqrt{3}}{2},$$

which can be achieved by either adjusting the angle θ of the same laser light or choosing three laser lights with different wavelengths λ_D. The resultant honeycomb lattice Hamiltonian is given by

$$H_s^{\text{hc}}(\mathbf{Z}) = -\frac{16}{9}\omega_0 \prod_{n=1}^{3} \sin^2\left(\frac{1}{2}\mathbf{v}_n \cdot \mathbf{Z}\right). \tag{5.2}$$

Here, we have defined the vector $\boldsymbol{Z} \equiv (X, P)$ and three ancillary vectors $v_1 = (1, \frac{1}{\sqrt{3}})$, $v_2 = (-1, \frac{1}{\sqrt{3}})$, $v_3 = (0, -\frac{2}{\sqrt{3}})$.

The two primitive vectors of honeycomb lattice's parallelogram primitive cell (unit cell with single atom) are $\boldsymbol{e}_1 = (2\pi, 0)$ and $\boldsymbol{e}_2 = (\pi, \sqrt{3}\pi)$, and the Bravais lattice is generated by

$$\boldsymbol{Z}_{l,s} = (X_{l,s}, P_{l,s}) = z_s + l_1\boldsymbol{e}_1 + l_2\boldsymbol{e}_2, \;\; l_1, l_2 \in \mathbb{Z}, \tag{5.3}$$

where $z_A = (0, 0)$ and $z_B = (\pi, \pi/\sqrt{3})$ are the origin points for the two triangular sublattices.

The two sublattices of the phase space honeycomb lattice Hamiltonian (5.2) have the same vibration frequency $\omega_0 = -9\Lambda$ at their lattice sites. One can also tune the difference of two sublattices on-site frequencies with the following phase space lattice Hamiltonian

$$H_s^{\Delta} = -\frac{2\Delta}{\sqrt{3}} \sum_{n=1}^{3} \sin\left(v_n \cdot \mathbf{Z}\right), \tag{5.4}$$

which can be implemented by additional stroboscopic lattices with kicking parameters $K_q = 2\Delta/\sqrt{3}$, $k_q = 2/\sqrt{3}$ and $\phi_q = 3\pi/2$ for $\theta_q = 1/12, 5/12, 9/12$ (see appendix A for more details). The on-site frequencies of two sublattices are then $\omega_0 \pm 2\Delta$ with the parameter Δ which we call the *on-site detuning*. The total single-particle Hamiltonian of honeycomb lattice including the on-site detuning terms is given by

$$H_s(\boldsymbol{Z}) = \frac{1}{2}\delta\omega|\boldsymbol{Z}|^2 - \frac{16}{9}\omega_0 \prod_{n=1}^{3} \sin^2\left(\frac{1}{2}\boldsymbol{v}_n \cdot \boldsymbol{Z}\right) - \frac{2\Delta}{\sqrt{3}} \sum_{n=1}^{3} \sin(\boldsymbol{v}_n \cdot \boldsymbol{Z}). \qquad (5.5)$$

Here, we have also introduced the detuning between kicking and harmonic oscillation $\delta\omega \equiv 1 - 2\pi/\tau$, which gives the first term on the right-hand side of the above equation. Note that Hamiltonian (5.5) has time-reversal symmetry of $H_s(X, P) = H_s(X, -P)$.

Quantum fluctuations

In quantum mechanics, we have the noncommutative relationship $[X, P] = i\lambda$ with the effective Planck constant $\lambda = \hbar k^2/m\omega_{ax}$. The condition $\lambda \ll 1$ ensures that the quantum fluctuations are small compared to the phase space distance between neighbouring sites. We estimate the nearest-neighbouring tunnelling rate given by the dominant tunnelling path as shown in figure 5.2(a). We expand the Hamiltonian along the dominant tunnelling path by

$$H_s^{\text{eff}} \approx \frac{P^2}{2m(X)} + V(X), \qquad (5.6)$$

where $V(X) = H_s(X, 0)$ has the form of double-well potential, and

$$m(X) = \left[\frac{\partial^2 H(X, P)}{\partial P^2}\right]^{-1} \qquad (5.7)$$

plays the role of effective mass. In figure 5.2(c), we plot $V(X)$ and the inverse of mass $1/m(X)$ along the dominant tunnelling path. For a given energy level E, the tunnelling rate can be calculated from the WKB approximation [12]

$$J = \frac{\lambda\omega_0}{2\pi} \exp\left(-\frac{1}{\lambda}\int_{-X_a}^{X_a} \left|\sqrt{2m(X)[E - V(X)]}\right| dX\right), \qquad (5.8)$$

where $\pm X_a$ is the turning point given by $V(X_a) = E$. In our case, the integral of action is divergent resulting a zero tunnelling rate. In fact, since the 2nd derivative $\partial^2 H(X, P)/\partial P^2$ vanishes at the saddle point of $V(X)$, the effective mass is divergent

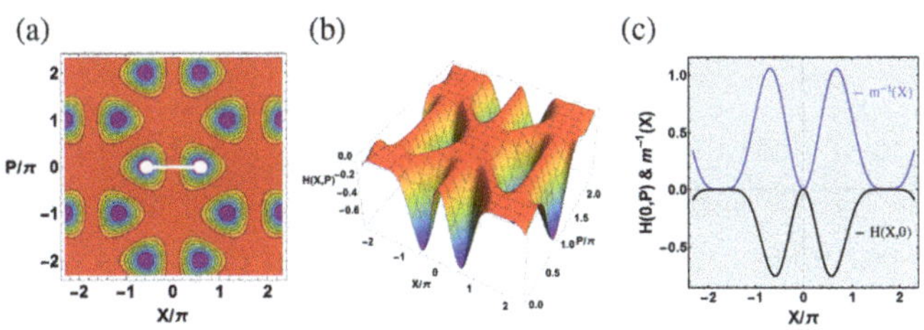

Figure 5.2. (a) Density plot of phase space honeycomb lattice Hamiltonian. The white line indicates the dominant tunnelling path. (b) 3D plot of phase space honeycomb lattice Hamiltonian showing a double-well structure along the line $P = 0$. (c) The zero momentum cut of Hamiltonian $H(X, 0)$ (black) and the inverse of effective mass $m(X)$ (blue). Produced using own data but following methods in [1].

at the saddle point. Therefore, the tunnelling rate is zero from the WKB approximation.

5.1.2 Many-body Hamiltonian

For the ultracold atom experiments, the collision of cold atoms is dominated by s-wave scattering process at low temperature. The two-body interactions of ultracold gases in 3D can be described by a pseudopotential in the form of contact function [3] $V_{3D}(\boldsymbol{r}_i - \boldsymbol{r}_j) = \frac{4\pi\hbar^2 a}{2m}\delta(\boldsymbol{r}_i - \boldsymbol{r}_j)$, where a is the 3D s-wave scattering length. The scattering length a can be further tuned by the Feshbach resonance with a magnetic field B, i.e., $a(B) = a_{\text{bg}}\left[1 - \frac{\Delta B}{B - B_0}\right]$. Here, a_{bg} is the off-resonant background scattering length, while ΔB and B_0 describe the width and position of the resonance. In the quasi-1D trap, the strength of contact interaction can be modified by the transverse mode with the effective pseudopotential is given by [3]

$$V_{1D}(x_i - x_j) = \frac{2\hbar\omega_{\text{tr}}a}{1 - Aa/l_{\text{tr}}}\delta(x_i - x_j) \approx 2\hbar\omega_{\text{tr}}a\delta(x_i - x_j), \tag{5.9}$$

where the constant $A = 1.036$ and $l_{\text{tr}} = \sqrt{\hbar/m\omega_{\text{tr}}}$ is the characteristic length of transverse motion. The approximation comes from the fact that the scattering length a is usually much shorter than the trapping length l_{tr}.

The interaction $V_{1D}(x_i - x_j)$ induces an effective interaction of two particles on their quadratures. In section 4.1.1, we have discussed the general method to extract this effective interaction in phase space. For the contact interaction $V_{1D}(x_i - x_j) = \gamma\delta(x_i - x_j)$, the effective interaction of two atoms on their quadratures is a Coulomb-like interaction given by

$$U(R_{ij}) = \frac{\gamma}{\pi}\frac{1}{R_{ij}}, \tag{5.10}$$

where $R_{ij} \equiv \sqrt{(X_i - X_j)^2 + (P_i - P_j)^2}$ is the phase space distance. The effective phase space interaction is valid for well-separated atoms, i.e., the lattice constant is much larger than the width of wave packet $d \gg \sqrt{\lambda}$ as discussed in section 4.2.2 and see equation (4.59). Finally, we have the many-body Hamiltonian [1]

$$H = \sum_L H_s(X_L, P_L) + \frac{1}{2}\sum_{L \neq L'} U(R_{LL'}) \equiv T + \Phi. \tag{5.11}$$

Here, we have relabelled the atoms by the subscript $L = (l, s)$ representing the lth atom in the $s \in \{A, B\}$ sublattice. $T \equiv \sum_L H_s(X_L, P_L)$ and $\Phi \equiv \frac{1}{2}\sum_{L \neq L'} U(R_{LL'})$ are the total single-particle part and the total interaction part respectively.

5.1.2.1 Equilibrium configuration

In the presence of the lattice potential and their effective interaction, the atoms have equilibrium configuration in phase space. The equilibrium point is determined by the condition

$$\frac{\partial H}{\partial u_L^\alpha} = 0, \quad \alpha \in \{X, P\}. \tag{5.12}$$

For the periodic boundaries in both X and P directions, the interactions from symmetric directions cancel each other and the equilibrium points are given by the honeycomb lattice sites, see equation (5.3). The periodic boundary is helpful for theoretical study. However, in the realistic experiments, the atoms tend to relax towards the stable points and concentrate about the origin of the phase space due to the presence of dissipation. By introducing a relative strong repulsive interaction ($\gamma > 0$), the atoms will spread over the phase space and form a disk-shaped crystal state, as shown by figure 5.3(a). Note that the interaction of the atoms on the disk will tend to distort the honeycomb equilibrium configuration. There is a mean-field potential induced by the interaction. The effective potential generated by the disk crystal in phase space can be approximated by the Coulomb potential on a uniformly charged disk plane (up to a constant due to the discretisation)

$$\bar{U}(\boldsymbol{Z}_i) = 2\pi\sigma R\left[1 - \sum_{l=0}(2l+1)\left[\frac{(2l-1)!!}{(2l+2)!!}\right]^2\left(\frac{|\boldsymbol{Z}_i|}{R}\right)^{2l+2}\right], \tag{5.13}$$

where R is the radius of disk, $|\boldsymbol{Z}_i| < R$ is the phase space distance to the original point of an atom on the disk, and $\sigma = \gamma/\sqrt{3}\pi^3$ is the effective charge density. The leading term ($l = 0$) in equation (5.13) produces a harmonic potential $\bar{U}(|\boldsymbol{Z}_i|) \sim -\sigma\pi R^{-1}|\boldsymbol{Z}_i|^2/2$. Considering contribution from high-order lth terms in equation (5.13), we find a better fitting with parabolic function gives

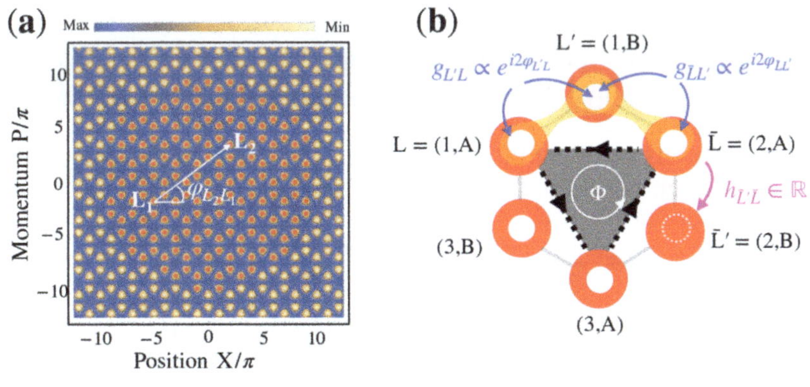

Figure 5.3. (a) Colour plot of single-particle Hamiltonian $H_s(X, P)$. The honeycomb lattice sites are occupied by atoms (red) forming a disk-shaped crystal in phase space. $\varphi_{L_2L_1}$ is the spatial angle of the vector connecting any two lattice sites (L_1 and L_2). (b) Hopping process (pink curved arrow) and pairing process (blue arrows and orange dumbbell-like shapes) of phase space lattice modes in one unit cell. The hopping rate $h_{\bar{L}'\bar{L}}$ is a real-valued number while the pairing rate $g_{L'L} \propto e^{i2\varphi_{L'L}}$ is a complex number with $\varphi_{L'L}$ being the spatial angle of the vectors connecting L and L' lattice sites, see (a) and equation (5.20). The black dashed lines represent the next-nearest-neighbour hopping ($\tilde{h}_{\bar{L}L} \propto e^{i\Phi/3}$) induced by two nearest-neighbouring pairing processes ($g_{L'L}$ and $g_{\bar{L}L'}$) with the arrows marking the directions of positive phase hopping, see equation (5.25). Produced using own data but following methods in [1].

$$\bar{U}(|Z_i|) \sim -1.35\sigma\pi R^{-1}|Z_i|^2/2. \tag{5.14}$$

Therefore, we can set the global detuning $\delta\omega \sim 1.35\sigma\pi R^{-1}$ in the Hamiltonian (5.5) to strongly counterbalance the interaction-induced mean-field potential.

5.1.2.2 Linearised lattice modes

We assume each atom oscillates around its equilibrium position, i.e., $\boldsymbol{Z}_L(t) = \boldsymbol{Z}_L^0 + \boldsymbol{u}_L(t)$ where footnote $L = (l, s)$ represents the lth atom in the $s \in \{A, B\}$ sublattice, and $\boldsymbol{u}_L = (u_L^X, u_L^P)$ is the displacement with respect to the equilibrium position $\boldsymbol{Z}_L^0$. The equilibrium points of the many-body Hamiltonian is determined by the condition (5.12). We expand the many-body Hamiltonian (5.11) around the equilibrium positions of atoms to the second order and obtain the linearised Hamiltonian [1] (up to a constant)

$$H = \frac{1}{2}\sum_{\alpha,\beta}\sum_{L,L'} \eta_{\alpha\beta}^{LL'} u_L^\alpha u_{L'}^\beta \quad \text{with} \quad \eta_{\alpha\beta}^{LL'} = \left.\frac{\partial^2 H}{\partial u_L^\alpha \partial u_{L'}^\beta}\right|_0 . \tag{5.15}$$

We call $\eta_{\alpha\beta}^{LL'}$ the *phase space force matrix*, which means the resultant force along the α direction exerted on the Lth atom due to the unit displacement along β direction of the L' th atom. From the definition of phase space force matrix, we have the property $\eta_{\alpha\beta}^{LL'} = \eta_{\beta\alpha}^{L'L}$.

We then introduce the complex field at each lattice site $a_L(t) \equiv [u_L^X(t) + iu_L^P(t)]/\sqrt{2\lambda}$, and obtain the Hamiltonian from equation (5.15)

$$\mathcal{H}/\lambda = \sum_{L,L'} h_{LL'} a_L^\dagger a_{L'} + \frac{1}{2} g_{LL'} a_L^\dagger a_{L'}^\dagger + \frac{1}{2} g_{L'L}^* a_L a_{L'} \tag{5.16}$$

Here, h_{LL} represents the on-site energy of one excitation on the Lth lattice site, $h_{LL'} (L \neq L')$ represents the hopping coefficient from the L' th lattice site to the Lth lattice site, and $g_{LL'}$ represents the pairing coefficient of creating or annihilating two phase space phonons on the Lth and L' th lattice sites. The hopping and pairing coefficients are given by the phase space force matrix

$$\begin{cases} h_{LL'} \equiv \frac{1}{2}\left(\eta_{XX}^{LL'} + \eta_{PP}^{LL'}\right) + i\frac{1}{2}\left(\eta_{PX}^{LL'} - \eta_{XP}^{LL'}\right) \\ g_{LL'} \equiv \frac{1}{2}\left(\eta_{XX}^{LL'} + \eta_{PP}^{LL'}\right) + i\frac{1}{2}\left(\eta_{PX}^{LL'} - \eta_{XP}^{LL'}\right) \end{cases}. \tag{5.17}$$

From the property $\eta_{\alpha\beta}^{LL'} = \eta_{\beta\alpha}^{L'L}$, we have the following important properties of hopping and pairing coefficients

$$h_{LL'} = h_{L'L}^*, \qquad g_{LL'} = g_{L'L}. \tag{5.18}$$

In appendix C, we have calculated the explicit forms of $h_{LL'}$ and $g_{LL'}$ for the case of Coulomb interaction equation (5.10). The explicit expressions of hopping and pairing coefficients are given as follows.

- For the off-site ($L \neq L'$) case, we have the hopping and pairing coefficients

$$h_{LL'} = -\frac{\gamma}{2\pi R_{LL'}^3}, \qquad g_{LL'} = -\frac{3\gamma}{2\pi R_{LL'}^3} e^{i2\varphi_{LL'}}. \tag{5.19}$$

  Here, the phase parameter $\varphi_{LL'}$ of the pairing coefficient is the spatial angle of the vectors connecting the L and L' lattice sites, see figure 5.3(a), defined via

$$X_L - X_{L'} = R_{LL'} \cos \varphi_{LL'}, \qquad P_L - P_{L'} = R_{LL'} \sin \varphi_{LL'}. \tag{5.20}$$

- For the on-site ($L = L'$) case, the hopping and pairing coefficients are the on-site energy $\omega_L \equiv h_{LL}$ and squeezing rate $g_L \equiv g_{LL}$ given by

$$\begin{cases} \omega_L = \dfrac{1}{2}(T_{XX}^{LL} + T_{PP}^{LL}) + \sum\limits_{L' \neq L} \dfrac{\gamma}{2\pi R_{LL'}^3} \\ g_L = \dfrac{1}{2}(T_{XX}^{LL} - T_{PP}^{LL}) + iT_{PX}^{LL} + \sum\limits_{L' \neq L} \dfrac{3\gamma}{2\pi R_{LL'}^3} e^{i2\varphi_{LL'}}. \end{cases} \tag{5.21}$$

  Here, $T_{\alpha\beta}^{LL} = \partial^2 H_s / \partial u_L^\alpha \partial u_L^\beta |_0$ is the phase space force matrix from the single-particle Hamiltonian.

Note that the expressions (5.19) and (5.21) are valid for any boundary condition and arbitrary single-particle Hamiltonian. For the honeycomb lattice Hamiltonian (5.5) with $\delta\omega = 0$ and periodic boundary conditions, we have the simplified results of on-site coefficients

$$\omega_L = \omega_0 \pm 2\Delta + \sum_{L' \neq L} \frac{\gamma}{2\pi R_{LL'}^3}, \quad g_L = 0, \tag{5.22}$$

where the summation in equation (5.21) disappears due to the honeycomb lattice symmetry. Finally, we obtain the many-body Hamiltonian for the linearised phase space crystal vibrations [1]

$$\begin{aligned} \frac{\mathcal{H}}{\lambda} &= \sum_L \omega_L a_L^\dagger a_L + \frac{1}{2} g_L a_L^{\dagger 2} + \frac{1}{2} g_L^* a_L^2 \\ &- \frac{\gamma}{4\pi} \left(\sum_{L \neq L'} \frac{a_L^\dagger a_{L'} + 3 e^{i2\varphi_{LL'}} a_L^\dagger a_{L'}^\dagger}{|\mathbf{Z}_L^0 - \mathbf{Z}_{L'}^0|^3} + h.\,c. \right). \end{aligned} \tag{5.23}$$

The Hamiltonian equation (5.23) describes phonons propagating in our honeycomb phase space crystal. It is reminiscent of the tight-binding model for electrons in graphene but our phase space phonons can hop between any two arbitrarily distant sites reflecting the long-range nature of the atom–atom interaction.

In our model, the hopping rates are always real-valued numbers $h_{LL'} \in \mathbb{R}$ while the pairing rates $g_{LL'} \propto e^{i2\varphi_{LL'}}$ have complex phase of $2\varphi_{LL'}$. As defined by equation (5.20) and shown in figure 5.3(a), the parameter $\varphi_{LL'}$, being the angle of the vector connecting two lattice sites in phase space, is not engineered by the driving field (independent of driving parameters) but comes from the intrinsic symplectic geometry of phase space together with phase space interaction $U(R_{ij})$. They are fundamentally different from the non-reciprocal phases of tight-binding (TB) Hamiltonians on a 2D real space lattice. This remains true for any arbitrary phase space Hamiltonian $H_s(\mathbf{Z})$ and atom–atom interaction $U(\mathbf{Z}_i - \mathbf{Z}_j)$, see appendix C.

5.1.3 Quantum anomalous Hall effect

In order to better distinguish the effects of the (long range) hopping and pairing, it is instructive to first consider a regime where the creation of phonon pairs is far off-resonant ($\gamma \ll \omega_0$). In this case, we can cancel the pairing-interaction terms in the Hamiltonian (5.16) using the following unitary transformation (see appendix C)

$$U \equiv \exp\left(-\sum_{\bar{L},\bar{L}'} \frac{g_{\bar{L}\bar{L}'}}{2\omega_{\bar{L}}} a_{\bar{L}}^{\dagger} a_{\bar{L}'}^{\dagger} + \frac{g_{\bar{L}\bar{L}'}^{*}}{2\omega_{\bar{L}}} a_{\bar{L}} a_{\bar{L}'}\right). \tag{5.24}$$

As a result, the pairing terms in the Hamiltonian (5.16) are cancelled and replaced by additional pairing-induced hopping terms

$$\tilde{h}_{LL'} \approx -\sum_{\bar{L}} \frac{2}{\omega_{\bar{L}}} g_{L\bar{L}} g_{\bar{L}L'}^{*}. \tag{5.25}$$

In the off-resonant regime, the main effect of pairing is to induce effective hopping transitions with non-reciprocal phases, see figure 5.3(b). Thus, a phonon propagating on a close loop $L(\cdot)$, e.g., a triangular loop enclosed by the dashed lines in figure 5.3(b), will acquire a geometrical Berry phase $\Phi(L(\cdot))$. These geometrical phases are similar to the Aharonov–Bohm phases acquired by charged particles in a magnetic field and are, thus, often referred to as synthetic magnetic fluxes. In fact, the effective particle-conserving description, which is derived from a perturbation theory in γ/ω_0, differs from the well-known Haldane model for the anomalous quantum Hall effect [13] only in that it includes long-range hopping transitions. We note that the phonon synthetic gauge field described here are not directly related to the synthetic gauge fields for quantum atoms tunnelling in the quasienergy surface $H_s(\mathbf{Z})$ [12, 14, 15].

Staggered magnetic field In order to illustrate how the pairing processes generate an artificial staggered magnetic field in phase space, we consider a toy model with consideration of symmetric honeycomb lattice ($\Delta = 0$) and only nearest-neighbour (NN) interaction, i.e., the interaction range for each atom is set to be $\delta = 2\pi/\sqrt{3}$. Using the labels of atoms shown in figure 5.3(b), and according to equations (5.22) and (5.21), we have the on-site hopping and pairing coefficients

$$\omega_L = \omega_0 + \frac{3\gamma}{2\pi\delta^3} = \omega_0 + \frac{9\sqrt{3}\gamma}{(2\pi)^4}, \quad g_L = 0 \tag{5.26}$$

and the off-site hopping and pairing coefficients

$$\begin{cases} h_{(1,A),(1,B)} = h_{(1,A),(2,B)} = -\dfrac{3\sqrt{3}}{(2\pi)^4}\gamma, \quad h_{(1,A),(3,B)} = 0, \\ g_{(1,A),(1,B)} = -\dfrac{9\sqrt{3}}{(2\pi)^4}\gamma e^{i\frac{\pi}{3}}, \; g_{(1,A),(2,B)} = -\dfrac{9\sqrt{3}}{(2\pi)^4}\gamma e^{i\frac{5\pi}{3}}, \quad g_{(1,A),(3,B)} = 0. \end{cases} \tag{5.27}$$

From equations (5.27) and (5.25), the NN pairings $g_{(1,A),(1,B)}$ and $g_{(1,A),(2,B)}$ induce hopping between the next-to-nearest-neighbour (NNN) lattice sites $L = (1, B)$ and $\bar{L} = (2, B)$,

$$\tilde{h}_{(1,B),(2,B)} = -\frac{2}{\omega_0} g_{(1,B),(1,A)}\, g^*_{(1,\,A),\,(2,\,B)} = -\frac{2}{\omega_0}\left[\frac{9\sqrt{3}}{(2\pi)^4}\right]^2 \gamma^2 \exp\left(i\frac{2\pi}{3}\right). \tag{5.28}$$

Here, we have adapted $\omega_L \approx \omega_0$ in the spirit of off-resonant condition ($\gamma/\omega_0 \ll 1$). It is not difficult to prove that the pairing-induced hopping $\tilde{h}_{(3,B),(1,B)}$ and $\tilde{h}_{(1,B),(3,B)}$ are equal to $\tilde{h}_{(1,B),(2,B)}$ from the view point of symmetry. Therefore, a phonon acquires a positive hopping phase $\Phi = 2\pi$ when it hops along a closed loop of the grey triangle as shown in figure 5.3(b). Note that the pairing-induced hopping is proportional to the square of interaction strength $|\tilde{h}_{L\bar{L}}| \propto \gamma^2/\omega_0$. Considering the case beyond NN interaction, there is a direct real-valued hopping rate between NNN lattice sites which is proportional to the interaction strength $h_{L\bar{L}} \propto \gamma$. As a result, the pairing-induced non-reciprocal hopping pathway $\tilde{h}_{L\bar{L}}$ interferes with a direct hopping pathway $h_{L\bar{L}}$ leads to a weak staggered magnetic flux of the order $\Phi \sim \gamma/\omega_0$. This is the essential ingredients of Haldane model [13], which introduces a staggered magnetic field breaking the time-reversal symmetry inside each unit cell but leaving the total net flux through the unit cell zero.

Our model recovers the anomalous quantum Hall effect similar to the Haldane model only in the weak off-resonant interaction regime. Our effective particle-conserving description (which is derived from a perturbation theory in γ/ω_0) differs from the Haldane model in that it includes long-range hopping transitions. For sufficiently strong interaction (non-perturbative regime), our model has some new features which do not appear in Haldane model, e.g., new topological phase regime and the loss of stability.

In the left panel of figure 5.4(a), we show the band structure in the weak interaction regime. There are two Dirac cones at the high-symmetry points K and K' with the tips of the cones separated by a small gap, which is the main distinguishing feature of its graphene counterpart although the long-range hopping strongly modifies it. Because the Dirac cones are symmetry-protected, the small band gap must be induced by the symmetry-breaking pairing-perturbation. The band gap must be topological as the broken symmetry is the local 'time-reversal symmetry' [13].

5.2 Topological phase space lattice waves

In this section, we discuss the general framework to diagonalise the linearised Hamiltonian (5.23) including the non-perturbative regime, where the bosonic

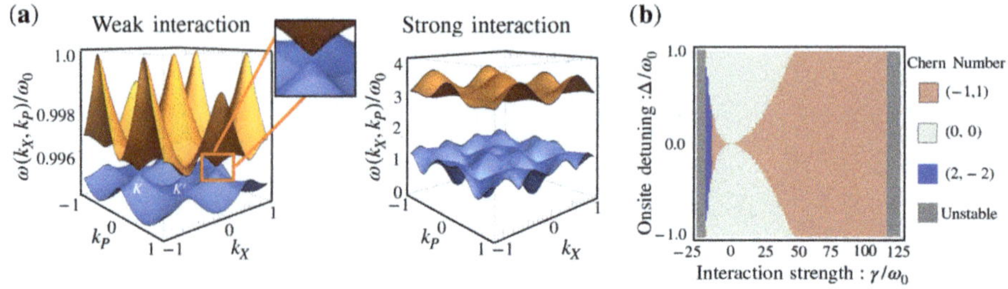

Figure 5.4. Bulk band structures of phase space lattice waves: (a) for weak interaction strength $\gamma/\omega_0 = -0.2$ (left) and strong interaction strength $\gamma/\omega_0 = 102$ (right). (c) Topological phase diagram with Chern number of bands as functions of interaction strength γ and on-site detuning Δ, which lifts the degeneracy of K and K' symmetry points. Produced using own data but following methods in [1].

pairing interaction γ is not a small perturbation compared to ω_0. This scenario goes beyond the standard topological settings of insulators and superconductors. Since, the phonon number is not even approximately conserved, it is not possible to employ a single-particle description. The physics is also different compared to superconducting systems (with fermionic pairing interactions). Here in our model, there is no limit to the occupation number of single-particle states, which can lead to the amplification of fluctuations and even to instabilities.

5.2.1 Bosonic Bogoliubov transformation

Our main task is to solve the eigen problem of Hamiltonian (5.23). We first discuss the case of phase space crystals with periodic boundaries both in X and P directions. In this setting, we introduce the ladder operators in the $\boldsymbol{k} = (k_X, k_P)$ reciprocal space by

$$a_{\boldsymbol{k},s} = \frac{1}{\sqrt{N}} \sum_l a_{l,s} \exp(-i\boldsymbol{k} \cdot \boldsymbol{Z}^0_{l,s}); \qquad a^\dagger_{\boldsymbol{k},s} = \frac{1}{\sqrt{N}} \sum_l a^\dagger_{l,s} \exp(i\boldsymbol{k} \cdot \boldsymbol{Z}^0_{l,s}), \quad (5.29)$$

where $\boldsymbol{Z}^0_{l,s}$ is the equilibrium position of the (l, s) th atom and N is the total number of primitive cells. By defining the ladder vector $\boldsymbol{A}_{\boldsymbol{k}} = (a_{\boldsymbol{k},A}, a_{\boldsymbol{k},B}, a^\dagger_{-\boldsymbol{k},A}, a^\dagger_{-\boldsymbol{k},B})^T$, we write Hamiltonian (5.16) in its Fourier transformation $\mathcal{H}/\lambda = \sum_{\boldsymbol{k}} \boldsymbol{A}^T_{\boldsymbol{k}} \mathcal{H}_{\boldsymbol{k}} \boldsymbol{A}_{\boldsymbol{k}}$, where the Fourier transformed Hamiltonian $\mathcal{H}_{\boldsymbol{k}}$ is given by (see appendix C)

$$\mathcal{H}_{\boldsymbol{k}} = \begin{pmatrix} h_{A,A}(\boldsymbol{k}) & h_{A,B}(\boldsymbol{k}) & \frac{1}{2}g_{A,A}(\boldsymbol{k}) & \frac{1}{2}g_{A,B}(\boldsymbol{k}) \\ h_{B,A}(\boldsymbol{k}) & h_{B,B}(\boldsymbol{k}) & \frac{1}{2}g_{B,A}(\boldsymbol{k}) & \frac{1}{2}g_{B,B}(\boldsymbol{k}) \\ \frac{1}{2}g^*_{A,A}(\boldsymbol{k}) & \frac{1}{2}g^*_{A,B}(\boldsymbol{k}) & h_{A,A}(-\boldsymbol{k}) & h_{A,B}(-\boldsymbol{k}) \\ \frac{1}{2}g^*_{B,A}(\boldsymbol{k}) & \frac{1}{2}g^*_{B,B}(\boldsymbol{k}) & h_{B,A}(-\boldsymbol{k}) & h_{B,B}(-\boldsymbol{k}) \end{pmatrix}. \quad (5.30)$$

Here, the matrix elements $h_{s,s'}(\boldsymbol{k})$ and $g_{s,s'}(\boldsymbol{k})$ are the hopping and pairing coefficients in the reciprocal space given by

$$\begin{cases} h_{s,s'}(\boldsymbol{k}) = \sum_l h_{(l,s),(l',s')} \exp[-i\boldsymbol{k} \cdot (\boldsymbol{Z}^0_{l,s} - \boldsymbol{Z}^0_{l',s'})] \\ g_{s,s'}(\boldsymbol{k}) = \sum_l g_{(l,s),(l',s')} \exp[-i\boldsymbol{k} \cdot (\boldsymbol{Z}^0_{l,s} - \boldsymbol{Z}^0_{l',s'})]. \end{cases} \tag{5.31}$$

From the discrete lattice translational symmetry of phase space crystals, we have the properties of $h^*_{s,s'}(\boldsymbol{k}) = h_{s',s}(\boldsymbol{k})$ and $g_{s,s'}(-\boldsymbol{k}) = g_{s',s}(\boldsymbol{k})$. Here, $\mathcal{H}_k$ is the *bosonic version of the Bogoliubov–de Gennes (BdG) Hamiltonian* in the study of superconductivity of electrons.

From the Heisenberg equation $da_{k,s}/dt = i\lambda^{-1}[\mathcal{H}, a_{k,s}]$, we derive the equation of motion (EOM) for the ladder vector, i.e., $dA_k/dt = -iD(\boldsymbol{k})A_k$ with the *dynamical matrix*

$$D(\boldsymbol{k}) = \begin{pmatrix} h_{A,A}(\boldsymbol{k}) & h_{A,B}(\boldsymbol{k}) & \bar{g}_{A,A}(\boldsymbol{k}) & \bar{g}_{A,B}(\boldsymbol{k}) \\ h_{B,A}(\boldsymbol{k}) & h_{B,B}(\boldsymbol{k}) & \bar{g}_{B,A}(\boldsymbol{k}) & \bar{g}_{B,B}(\boldsymbol{k}) \\ -\bar{g}^*_{A,A}(\boldsymbol{k}) & -\bar{g}^*_{B,A}(\boldsymbol{k}) & -h^*_{A,A}(-\boldsymbol{k}) & -h^*_{A,B}(-\boldsymbol{k}) \\ -\bar{g}^*_{A,B}(\boldsymbol{k}) & -\bar{g}^*_{B,B}(\boldsymbol{k}) & -h^*_{B,A}(-\boldsymbol{k}) & -h^*_{B,B}(-\boldsymbol{k}) \end{pmatrix} \tag{5.32}$$

where $\bar{g}_{s,s'}(\boldsymbol{k}) \equiv \frac{1}{2}\left[g_{s,s'}(\boldsymbol{k}) + g_{s',s}(-\boldsymbol{k})\right]$ is the symmetrised pairing coefficient. By diagonalising the dynamical matrix $D(\boldsymbol{k})$, we obtain four solutions (eigenfrequency) for each $\boldsymbol{k}$. We label the frequency spectrum by $\omega_{k,n}$, where $n = 1, 2, 3, 4$ is the index of four bands. For each given $\boldsymbol{k}$ and band index n, the eigenstate $|\boldsymbol{k}, n\rangle = (u_{n,A}, u_{n,B}, v_{n,A}, v_{n,B})$ is a four-component vector. In this way, we equivalently cast the Hamiltonian (5.16) into a diagonal form of

$$\mathcal{H}/\lambda = \sum_{k,n} \omega_{k,n} b^\dagger_{k,n} b_{k,n}, \tag{5.33}$$

where the normal modes are given by a linear transformation in the form of

$$b^\dagger_{k,n} = \sum_{s=A,B} u_{n,s}(\boldsymbol{k}) a^\dagger_{k,s} + v_{n,s}(\boldsymbol{k}) a_{-k,s}. \tag{5.34}$$

From the property of symmetrised pairing coefficient $\bar{g}_{s,s'}(\boldsymbol{k}) = \bar{g}_{s,s'}(-\boldsymbol{k})$, we find the dynamical matrix has *particle–hole symmetry* expressed by

$$\Xi D(\boldsymbol{k}) \Xi^{-1} = -D(-\boldsymbol{k}). \tag{5.35}$$

The particle–hole operator is defined via $\Xi = \tau_x K$ satisfies $\Xi^2 = +1$, where

$$\tau_x = \begin{pmatrix} 0 & 0 & 1 & 0 \\ 0 & 0 & 0 & 1 \\ 1 & 0 & 0 & 0 \\ 0 & 1 & 0 & 0 \end{pmatrix} \tag{5.36}$$

and K is the complex conjugation. It follows that each eigenmode at frequency $\omega(\boldsymbol{k})$ has a partner eigenmode at $-\omega(-\boldsymbol{k})$, namely, creating a quasiparticle in the state $\omega(\boldsymbol{k})$ has the same effect as removing one (creating a hole) from the state $-\omega(-\boldsymbol{k})$. Therefore, we label the two upper bands ($n = 1, 2$) as particle bands and two lower bands ($n = 3, 4$) as the hole bands. From the bosonic commutation relationship $[b_{\boldsymbol{k}',n'}, b^{\dagger}_{\boldsymbol{k},n}] = \delta_{n,n'}\delta_{\boldsymbol{k}',\boldsymbol{k}}$, the eigenmodes have to follow the ortho-normal condition

$$\langle \boldsymbol{k}, n'|\Sigma_z|\boldsymbol{k}, n\rangle = \sum_s u^*_{n',s}u_{n,s} - v^*_{n',s}v_{n,s} = \pm\delta_{n,n'}, \tag{5.37}$$

where $\Sigma_z \equiv \text{diag}(1, 1, -1, -1)$ is a 4×4 diagonal matrix. The positive (negative) sign in equation (5.37) corresponds to particle (hole) bands. The transformation (5.34) with constraint of ortho-normal condition (5.37) is called the *bosonic Bogoliubov transformation.*

5.2.2 Topological bulk bands

As discussed above, due to the non-reciprocal phase $\varphi_{LL'}$, the bands of Hamiltonian (5.23) should have non-trivial topological structure in the quasiparticle wavefunctions $|\boldsymbol{k}, n\rangle$. To quantify this, we first need to properly update the definition of Berry phase and Chern number associated with a bosonic band eigenstate in the presence of pairing terms that violate the particle number conservation.

5.2.2.1 Symplectic Chern number

Different from the particle-conserving case, the ground state of the BdG Hamiltonian $\mathcal{H}_{\boldsymbol{k}}$ is a multi-mode squeezed state with non-zero photon number. By regarding the $\boldsymbol{k} = (k_X, k_P)$ as an external adiabatic parameter, we calculate the Berry phase accumulated by a single Bogoliubov quasiparticle in a specific nth band along a closed loop covering the whole Brillouin zone (BZ), i.e.,

$$\begin{aligned}
\Phi_n &= i\oint_{\text{BZ}} \langle S_{\boldsymbol{k}}|b_{\boldsymbol{k},n}\nabla_{\boldsymbol{k}}b^{\dagger}_{\boldsymbol{k},n}|S_{\boldsymbol{k}}\rangle \cdot d\boldsymbol{k} \\
&= i\oint_{\text{BZ}} \langle S_{\boldsymbol{k}}|[b_{\boldsymbol{k},n}\nabla_{\boldsymbol{k}}b^{\dagger}_{\boldsymbol{k},n}]|S_{\boldsymbol{k}}\rangle \cdot d\boldsymbol{k} + i\oint_{\text{BZ}} \langle S_{\boldsymbol{k}}|\nabla_{\boldsymbol{k}}|S_{\boldsymbol{k}}\rangle \cdot d\boldsymbol{k} \\
&= i\oint_{\text{BZ}} \left(\sum_s u^*_{n,s}(\boldsymbol{k})\nabla_{\boldsymbol{k}}u_{n,s}(\boldsymbol{k}) - v^*_{n,s}(\boldsymbol{k})\nabla_{\boldsymbol{k}}v_{n,s}(\boldsymbol{k})\right) \cdot d\boldsymbol{k} \\
&\qquad\qquad + i\oint_{\text{BZ}} \langle S_{\boldsymbol{k}}|\nabla_{\boldsymbol{k}}|S_{\boldsymbol{k}}\rangle \cdot d\boldsymbol{k} \\
&\equiv \oint_{\text{BZ}} \mathcal{A}_n(\boldsymbol{k}) \cdot d\boldsymbol{k} + \phi_n.
\end{aligned} \tag{5.38}$$

Here, $|S_{\boldsymbol{k}}\rangle$ is the ground state (Bogoliubov vacuum state) of $\mathcal{H}_{\boldsymbol{k}}$ and, in the second line, we have used $b_{\boldsymbol{k},n}|S_{\boldsymbol{k}}\rangle = 0$ by the definition of vacuum state. Because of the unusual ortho-normalisation (5.37), we have updated the definition of the Berry connection [16]

$$\mathcal{A}_n(\boldsymbol{k}) = i\langle \boldsymbol{k}, n|\Sigma_z \nabla_k|\boldsymbol{k}, n\rangle. \tag{5.39}$$

Note that the Berry connection defined here for the bosonic many-body second-quantised Hamiltonian is different from the Berry connection for the single-particle Hamiltonian given by equation (3.78) in section 3.3.1.

The Bogoliubov vacuum $|S_k\rangle$ depends on $\boldsymbol{k}$ and could possibly accumulate a Berry phase, i.e., $\phi_n \equiv i\oint \langle S_k|\nabla_k|S_k\rangle \cdot d\boldsymbol{k} \neq 0$. However, the Berry phase of our interest is the additional Berry phase accumulated by the quasiparticle, i.e., the difference of Bogoliubov vacuum Berry phase and that associated with a single quasiparticle excitation. We thus update the definition of Chern number by [16]

$$C_n = \frac{1}{2\pi}\int_{\mathrm{BZ}} (\nabla_k \times \mathcal{A}_n(\boldsymbol{k})) \cdot \hat{\boldsymbol{n}} \in \mathbb{Z}, \tag{5.40}$$

where $\hat{\boldsymbol{n}}$ is the unit vector normal to the $\boldsymbol{k}$-plane. Due to the additional element Σ_k in the Berry connection (5.39), the quantity given by equation (5.40) is also called the *symplectic Chern number*. Since the quantum states come back to itself along a closed loop, the quantities Φ_n and ϕ_n are integers multiple of 2π, and thus the symplectic Chern number is also an integer. As discussed in section 5.1.3, there is no net geometric phase (total flux of synthetic magnetic field) in our model, the sum of the Chern numbers over the particle bands must be zero. The Chern number of individual band may change after a phase transition where two or more bands touch each other, but their sum does not change. For this reason, we define $C \equiv |C_n|$ as the Chern number of our system.

5.2.2.2 Topological phase diagram

In figure 5.4(b), we show the topological phase diagram of our model by plotting the Chern number of bands as functions of the interaction strength γ/ω_0 and the on-site detuning Δ/ω_0, which breaks the inversion symmetry opening the possibility of a trivial band gap [13]. For $\Delta = 0$, the phase crystal has Chern number $C = 1$ for a broad range of attractive and repulsive interactions, but can also switch to $C = 2$ for moderate interactions $-20 \lesssim \gamma/\omega_0 \lesssim -15$. The $C = 2$ topological phase is not present in the Haldane model for the anomalous quantum Hall effect [13]. In the region of weak interactions, the band edge is coned-shaped and the band gap scales as γ^2/ω_0 for $\Delta = 0$, the latter quantity determines how far the topological region extends into the $\Delta \neq 0$ region.

For large attractive interaction $\gamma < \gamma_- \approx -20\omega_0$ or repulsive interaction $\gamma > \gamma_c^+ \approx 115\omega_0$, the phase space crystal becomes unstable as marked by the grey regions in the phase diagram. The bulk band structure for a comparatively strong interaction is shown in the right subfigure of figure 5.4(a). In the presence of bosonic pairing interactions, the quasienergy can be viewed as the energy cost of producing a pair of Bogoliubov excitations. When this hits zero for a critical threshold γ_+ or γ_-, where the band structure at the K and K'-points approaches zero quasienergy, the phase space crystal becomes unstable leading to a disordered gas-like phase in phase space.

5.2.3 Chiral edge states

In this section, we calculate the band structures for two different open boundary conditions, i.e., the strip-shaped boundary and the disk-shaped boundary. We show the topological edge states are chiral and robust against defects on the boundary. The chirality of topological edge states in our models keeps invariant under the non-local time-reversal transformation.

5.2.3.1 Strip-shaped boundary

We now consider the system with periodical boundary in the X direction but open boundaries in the P direction (i.e., strip-shaped boundary as shown in figure 5.5(b)). In this case, we can only perform the Fourier transformation in the X direction

$$a_{l,s} = \frac{1}{\sqrt{N_X}} \sum_{k_X} a_{k_X,(l_P,s)} \exp(ik_X X^0_{l,s}); \quad a^\dagger_{l,s} = \frac{1}{\sqrt{N_X}} \sum_{k_X} a^\dagger_{k_X,(l_P,s)} \exp(-ik_X X^0_{l,s}), \tag{5.41}$$

where N_X is the number of unit cells in the X direction and (l_P, s) labels the position of atoms in the P direction. By defining the ladder vector operator

$$\mathbf{A}_{l_p}(k_X) \equiv (a_{k_X,(l_P,A)}, a_{k_X,(l_P,B)}, a^\dagger_{-k_X,(l_P,A)}, a^\dagger_{-k_X,(l_P,B)})^T,$$

we have the EOM from Heisenberg's equation (see details in appendix C.4)

$$i\frac{d}{dt}\begin{pmatrix} \mathbf{A}_1 \\ \vdots \\ \mathbf{A}_{l'_P} \\ \vdots \\ \mathbf{A}_{N_P} \end{pmatrix} = \begin{pmatrix} D_{(l_P=1,l'_P=1)} & \cdots & D_{(l_P=1,l'_P)} & \cdots & D_{(l_P=1,l'_P=N_P)} \\ \vdots & \ddots & \vdots & \vdots & \vdots \\ D_{(l_P,l'_P=1)} & \cdots & D_{l'_P,l_P} & \cdots & D_{(l_P,l'_P=N_P)} \\ \vdots & \vdots & \vdots & \ddots & \vdots \\ D_{(l_P=N_P,l'_P=1)} & \cdots & D_{(l_P=N_P,l'_P)} & \cdots & D_{(l_P=N_P,l'_P=N_P)} \end{pmatrix}\begin{pmatrix} \mathbf{A}_1 \\ \vdots \\ \mathbf{A}_{l'_P} \\ \vdots \\ \mathbf{A}_{N_P} \end{pmatrix},$$

where the block matrix element is given by $D_{(l_P,l'_P)}(k_X) \equiv \Sigma_z \mathcal{H}_{(l_P,l'_P)}(k_X)$ with

$$\mathcal{H}_{(l_P,l'_P)}(k_X) =$$
$$\begin{pmatrix} h_{(l_P,A),(l'_P,A)}(k_X) & h_{(l_P,A),(l'_P,B)}(k_X) & \bar{g}_{(l_P,A),(l'_P,A)}(k_X) & \bar{g}_{(l_P,A),(l'_P,B)}(k_X) \\ h_{(l_P,B),(l'_P,A)}(k_X) & h_{(l_P,B),(l'_P,B)}(k_X) & \bar{g}_{(l_P,B),(l'_P,A)}(k_X) & \bar{g}_{(l_P,B),(l'_P,B)}(k_X) \\ \bar{g}^*_{(l_P,A),(l'_P,A)}(-k_X) & \bar{g}^*_{(l_P,A),(l'_P,B)}(-k_X) & h^*_{(l_P,A),(l'_P,A)}(-k_X) & h^*_{(l_P,A),(l'_P,B)}(-k_X) \\ \bar{g}^*_{(l_P,B),(l'_P,A)}(-k_X) & \bar{g}^*_{(l_P,B),(l'_P,B)}(-k_X) & h^*_{(l_P,B),(l'_P,A)}(-k_X) & h^*_{(l_P,B),(l'_P,B)}(-k_X) \end{pmatrix}.$$

Here, we have introduced the modified pairing coefficient

$$\bar{g}_{(l_P,s),(l'_P,s')}(k_X) \equiv \frac{1}{2}\left[g_{(l_P,s),(l'_P,s')}(k_X) + g_{(l'_P,s'),(l_P,s)}(-k_X)\right], \tag{5.42}$$

and the Fourier transformations of hopping terms and pairing coefficients

$$\begin{cases} h_{(l_P,s),(l'_P,s')}(k_X) = \sum_{l_X} h_{(l,s),(l',s')} \exp[-ik_X \cdot (X^0_{l,s} - X^0_{l',s'})] \\ g_{(l_P,s),(l'_P,s')}(k_X) = \sum_{l_X} g_{(l,s),(l',s')}(l - l') \exp[-ik_X \cdot (X^0_{l,s} - X^0_{l',s'})]. \end{cases} \tag{5.43}$$

From the discrete translational symmetry in the X direction and equation (5.18), we have the following properties of hopping and pairing coefficients in reciprocal space

$$h^*_{(l'_P,s'),(l_P,s)}(k_X) = h_{(l_P,s),(l'_P,s')}(k_X), \qquad g_{(l_P,s),(l'_P,s')}(-k_X) = g_{(l'_P,s'),(l_P,s)}(k_X). \tag{5.44}$$

By extending the definition of operator τ_x, see equation (5.36), to all the matrix elements of $D(k_X)$ labelled by (l_P, l'_P) and using the property $\bar{g}_{(l_P,s),(l'_P,s')}(k_X) = \bar{g}_{(l_P,s),(l'_P,s')}(-k_X)$, the dynamical matrix $D_{(l_P,l'_P)}(k_X)$ has *particle–hole symmetry* expressed by

$$\Xi D_{(l_P,l'_P)}(k_X)\Xi^{-1} = -D_{(l_P,l'_P)}(-k_X), \tag{5.45}$$

where the particle–hole operator is defined via $\Xi = \tau_x K$ satisfies $\Xi^2 = +1$. Again, we obtain the eigenmodes by diagonalising the dynamical matrix $D(k_X)$.

We plot the resultant band structure in figure 5.5(a) and the topological edge states in figure 5.5(b). We only plot the positive spectrum of particles since the negative spectrum of holes is central symmetrical with respect to k_X due to the particle–hole symmetry of dynamical matrix. It is clear that the edge states are chiral, i.e., that on the positive (negative) P boundary propagates to the left (right). In the plots, we have assumed the honeycomb equilibrium configuration is not distorted by the interaction, which is the conventional way to analyse the properties of the edge states. In figure 5.5(b), we also display the physical orbits of atoms near the stable points (exaggerated for vision), which shows the atoms move on elliptical trajectories with different phases of motion.

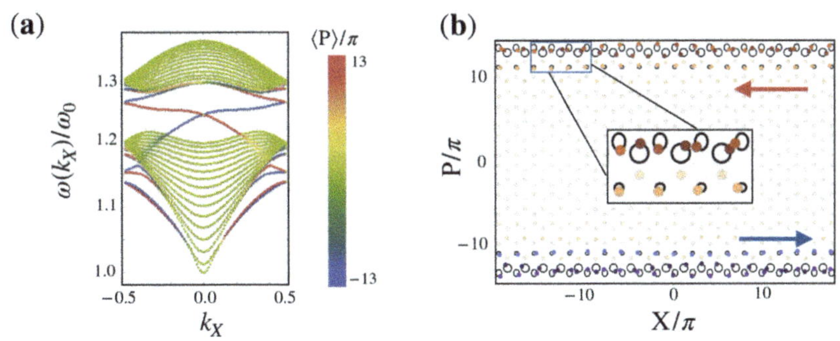

Figure 5.5. Topological edge states. (a) Band structure of phase space crystal with strip-shaped boundary (open boundaries in the P direction), where the colour indicates the average P value of each eigenstate. (b) Two topological edge states propagating on the P boundaries, where the circles indicate atoms' orbits and the colour indicates the intensity of lattice waves. Produced using own data but following methods in [1].

The number of topological edge states on one single boundary is concluded by the bulk-boundary correspondence, i.e., the difference between the net number of chiral edge states (counted with a sign depending on their chirality) at two energies is equal to the total Chern number of the intermediate bands. In our model, there are no topological edge states near the zero energy since the particle band cannot touch the hole band for the stable phase space crystals. Therefore, the Chern numbers of bands are sufficient to determine the number of topological edge states [17].

5.2.3.2 Disk-shaped boundary

In figure 5.6(a), we consider a disk-shaped crystal with some random defects on the boundary. Here, we take fully into account the atom–atom interaction including also the deformation of the equilibrium configuration. As discussed above, we reduce this effect using a laser detuning $\delta\omega = 1.35\sigma\pi R^{-1}$, see equation (5.14). Since there is no way to simplify the calculation by Fourier transformation for this case, we obtain the spectrum and the eigenstates by directly diagonalising the Hamiltonian (5.16).

The eigenstates $\{\phi_m\}$ are sorted by ascending order as shown in figure 5.6(b). In order to show the time evolution of topological transport on the edge, we choose the initial state as a Gaussian wave packet

$$\Psi(\boldsymbol{Z}_i, t = 0) \propto \sum_m \exp[-\frac{(m - m_c)^2}{2\sigma^2} + i\varphi_m(\boldsymbol{Z}_c)]\phi_m(\boldsymbol{Z}_i), \tag{5.46}$$

with average quasienergy ω_c in the middle of the band gap, as indicated by the arrow in figure 5.6(b). Numerically, the centre of wave packet can be positioned by gauging all the eigenstates as real number on the chosen lattice site, i.e., by choosing the additional phases $\varphi_m(\boldsymbol{Z}_i)$ in equation (5.46) to cancel the uncertain complex phases of $\phi_m(\boldsymbol{Z}_i)$ at the position $\boldsymbol{Z}_i = \boldsymbol{Z}_c$ in phase space. We plot the wave packet at two

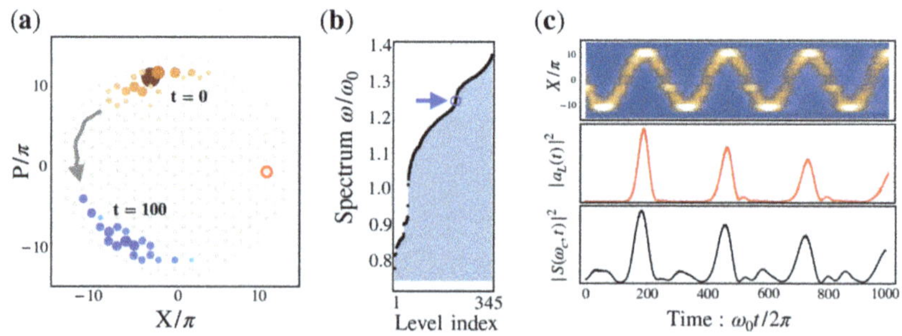

Figure 5.6. Topological edge states. (a) Disk-shaped phase space crystal with random defects on the edge with two topological wave packets at different moment instants. (b) Energy spectrum with ascending order. The wave packet in (a) is constructed by the Gaussian superposition of eigenstates centred at the energy level indicated by the blue arrows, see equation (5.46). (c) Spacetime plot (upper panel) of topological wave packet projected on the spatial X dimension, vibration $|a_L(t)|^2$ of the atom marked by the red circle in (a) (middle panel), and finite-window power spectrum $S(\omega_c, t)$ of all the atoms (lower panel). Parameters: interaction strength $\gamma = 25\omega_0$, disk radius $R = 15\pi$. Produced using own data but following methods in [1].

different time instants showing that it is robust against the defects on the edge during the propagation.

Considering our model is actually a 1D physical system in real space, we project the propagating wave packet $\Psi(X_i, P_i, t)$ on the spatial dimension and define the smoothed 1D intensity distribution

$$\rho(X, t) = \sum_{X_i \in \sigma} |\Psi(X_i, P_i, t)|^2. \tag{5.47}$$

In our calculation, we use $\sigma = [X - \pi, X + \pi]$ to get the smoothed projected 1D spatial density distribution. In the upper panel of figure 5.6(c), we show the spacetime plot of $\rho(X, t)$. It is clear that the topological wave packet is robust against the defects on the edge during the propagation. In the middle panel of figure 5.6(c), we plot the vibration of a fixed atom on the disk edge as marked by the red circle in figure 5.6(a).

In the real experiment, it may be difficult to measure the vibration of a single atom. Instead, one can detect the atoms' collective motion $I(t) = \sum_j \cos[k_s x_j(t)]$ via additional probe stroboscopic lattice and measure the finite-time-window power spectrum

$$S(\omega, t) \propto |\langle (I(t) - \langle I \rangle) e^{i\omega t} \rangle_{\Delta t}|^2, \tag{5.48}$$

where $\langle \cdots \rangle_{\Delta t}$ represents the time average over a finite-time widow. In our simulation, we choose the time window $\Delta t = 10 \times 2\pi\omega_0^{-1}$. In the lower panel of figure 5.6(c), we plot the power spectrum $S(\omega, t)$ at the wave packet central frequency $\omega = \omega_c$ and $k_s = 9.8$ as a function of time. It is clear that the peaks of power spectrum correspond to the chiral motion of topological wave packet. In the experimental parameters with ultracold ^{87}Rb atoms. Assuming a typical longitudinal trapping frequency $\omega_{\rm ax} = 2\pi \times 50$ Hz, we take $k^{-1} = 45\mu$m to work in the semi-classical regime ($\lambda = 0.04$). In the experiment, the transverse trapping frequency can reach up to $\omega_{\rm tr} \approx 2\pi \times 1.0$ MHz [3]. To avoid exciting the transverse mode during collision, there is a restriction for radius of phase space crystal

$$2 \times \frac{1}{2} m\omega_{\rm ax}^2 (k^{-1}R)^2 < \hbar\omega_{\rm tr} \quad \Longrightarrow \quad R < k\sqrt{\frac{\hbar\omega_{\rm tr}}{m\omega_{\rm ax}^2}} \approx 9.2\pi. \tag{5.49}$$

This radius is roughly half of the radius of disk shown in figure 5.6(a). By choosing $\omega_0 = 0.1$ (to guarantee slow vibrations in the rotating frame), the wave packet's travelling time of a round-trip along the disk edge would be approximately $2\pi \times 100/(\omega_{\rm ax}\omega_0) \approx 20$ s, shorter than the typical lifetime of the atom cloud [18–20]. Taking into account the 3D scattering length a= 5.3nm for ^{87}Rb atoms, the 1D dimensionless interaction strength is $\gamma = 2\hbar\omega_{\rm tr} a k^3/(m\omega_{\rm ax}^2) \approx 1.25$, which can be further tuned by Feshbach resonance or adjusting k, see section 5.1.2.

5.2.4 Further discussions

5.2.4.1 Time-reversal symmetry and chirality

In 2D real space, time-reversal transformation changes the chirality of trajectories, e.g., from clockwise to anti-clockwise. Therefore, robust chiral edge states in 2D real space can only be implemented after breaking the time-reversal symmetry without a time-reversed partner with opposite chirality. However, this constraint does not apply to phase space crystals, i.e., the chirality of motion in phase space remains unchanged under time-reversal transformation. On one hand, the many-body Hamiltonian (5.11) keeps invariant under the time-reversal transformation $(X_i \to X_i, P_i \to -P_i)$ which rearranges the phase space lattice in a non-local fashion. On the other hand, the effective Hamiltonian (5.23) does not support any local anti-unitary symmetry due to the complex phases $\varphi_{LL'}$, which is in fact the standard formal precondition for non-trivial Chern numbers and chiral edge states.

5.2.4.2 Comparison with synthetic dimensions

Our study is closely related to the method with synthetic dimensions [21]. The central idea of synthetic dimensions is to exploit and harness some of the internal degrees of freedom of atoms or photons to simulate motion along extra spatial directions. In our model, the P dimension may be viewed as a 'synthetic' dimension. However, strictly speaking, the P dimension is not another independent degree of freedom of system but the conjugate dimension of the X dimension. The hopping phases come from the combination of symplectic dynamics and interactions in phase space. Furthermore, it is unlikely to create a honeycomb lattice and the disk-shaped geometry in synthetic dimensions. Most of the known schemes have exotic inter-particle interactions and only discrete degrees of freedom along the synthetic direction, which hinders the simulation of models with local interactions or with continuous dimensions [21]. Our model provides an alternative way to create 2D topological phenomena with 1D physical system in parallel with the synthetic dimension methods.

5.2.4.3 Realisation in real space

Another interesting connection is that our model can be mapped into the magnetron motion of charged particles in 2D real space. The dynamics of a charged particle is described by Lorentz EOM

$$m\ddot{\boldsymbol{r}}_i = \boldsymbol{F}_i + q\dot{\boldsymbol{r}}_i \times \boldsymbol{B},$$

where $\boldsymbol{F}_i = -q\nabla_i\phi_s$ is the electrical force and $\boldsymbol{B} = B\hat{n}$ is the magnetic field perpendicular to the 2D plane. As discussed in section 4.1.2, the total trajectory of the charged particle can be decomposed into the cyclotron motion and the magnetron motion (i.e., drifted dynamics of the cyclotron centre). In the limit of infinity large charge-to-mass ratio $q/m \to \infty$, the dynamics of magnetron motion is

$$\dot{\boldsymbol{r}}_i = (qB)^{-1}\boldsymbol{F}_i \times \hat{n},$$

which is the same as the 2nd Newton's law in phase space, see equation (4.23). Considering phase space interaction $U(R_{ij})$ in equation (5.10) has the Coulomb form, the conclusions from our model can be extended directly to the magnetron motion of charged particles in the electrical potential $\phi_s \propto H_s$.

Note added: The work in chapter 5 is basically from the preprint arXiv:2105.06989, which has been submitted and under review. We cannot make change to the cited reference [1] after this book is published. But we refer the readers to the updated publication information about the work on the website.

References

[1] Guo L, Peano V and Marquardt F 2021 Phase space crystal vibrations: chiral edge states with preserved time-reversal symmetry arXiv:2105.06989 (submitted and under review, see the updated publication information)

[2] Paredes B, Widera A, Murg V, Mandel O, Fölling S, Cirac I, Shlyapnikov G V, Hänsch T W and Bloch I 2004 Tonks-Girardeau gas of ultracold atoms in an optical lattice *Nature* **429** 277–81

[3] Bloch I, Dalibard J and Zwerger W 2008 Many-body physics with ultracold gases *Rev. Mod. Phys.* **80** 885–964

[4] Moritz H, Stöferle T, Köhl M and Esslinger T 2003 Exciting collective oscillations in a trapped 1d gas *Phys. Rev. Lett.* **91** 250402

[5] Hadzibabic Z, Stock S, Battelier B, Bretin V and Dalibard J 2004 Interference of an array of independent Bose-Einstein condensates *Phys. Rev. Lett.* **93** 180403

[6] Zaslavsky G M, Zakharov M Y, Sagdeev R Z, Usikov D A and Chernikov A A 1986 Stochastic web and diffusion of particles in a magnetic field *Zh. Eksp. Teor. Fiz.* **91** 500

[7] Moore F L, Robinson J C, Bharucha C F, Sundaram B and Raizen M G 1995 Atom optics realization of the quantum δ-kicked rotor *Phys. Rev. Lett.* **75** 4598–601

[8] Chabé J, Lemarié G, Grémaud B, Delande D, Szriftgiser P and Garreau J C 2008 Experimental observation of the Anderson metal-insulator transition with atomic matter waves *Phys. Rev. Lett.* **101** 255702

[9] Manai I, Clément J-F, Chicireanu R, Hainaut C, Garreau J C, Szriftgiser P and Delande D 2015 Experimental observation of two-dimensional Anderson localization with the atomic kicked rotor *Phys. Rev. Lett.* **115** 240603

[10] Lemos G B, Gomes R M, Walborn S P, Souto Ribeiro P H and Toscano F 2012 Experimental observation of quantum chaos in a beam of light *Nat. Commun.* **3** 1211

[11] Zaslavsky G M 2008 *Hamiltonian Chaos and Fractional Dynamics* (Oxford: Oxford University Press)

[12] Guo L, Marthaler M and Schön G 2013 Phase space crystals: a new way to create a quasienergy band structure *Phys. Rev. Lett.* **111** 205303

[13] Haldane F D M 1988 Model for a quantum hall effect without landau levels: condensed-matter realization of the 'parity anomaly *Phys. Rev. Lett.* **61** 2015–8

[14] Zhang Y, Gosner J, Girvin S M, Ankerhold J and Dykman M I 2017 Time-translation-symmetry breaking in a driven oscillator: from the quantum coherent to the incoherent regime *Phys. Rev.* A **96** 052124

[15] Liang P, Marthaler M and Guo L 2018 Floquet many-body engineering: topology and many-body physics in phase space lattices *New J. Phys.* **20** 023043

[16] Peano V, Houde M, Brendel C, Marquardt F and Clerk A A 2016 Topological phase transitions and chiral inelastic transport induced by the squeezing of light *Nat. Commun.* **7** 10779

[17] Rudner M S, Lindner N H, Berg E and Levin M 2013 Anomalous edge states and the bulk-edge correspondence for periodically driven two-dimensional systems *Phys. Rev.* X **3** 031005

[18] Davis K B, Mewes M O, Andrews M R, van Druten N J, Durfee D S, Kurn D M and Ketterle W 1995 Bose-Einstein condensation in a gas of sodium atoms *Phys. Rev. Lett.* **75** 3969–73

[19] Greif D G 2007 Evaporative cooling and Bose–Einstein condensation of Rb-87 in a moving-coil top trap geometry *Master thesis*

[20] Cornell E A and Wieman C E 2002 Nobel lecture: Bose-Einstein condensation in a dilute gas, the first 70 years and some recent experiments *Rev. Mod. Phys.* **74** 875–93

[21] Ozawa T and Price H M 2019 Topological quantum matter in synthetic dimensions *Nat. Rev. Phys.* **1** 349–57

Chapter 6

Time crystals

In this chapter, we discuss a different but closely related topic: *time crystals*. Both phase space crystals and time crystals study subharmonic modes of periodically driven systems. Time crystals focus on the spontaneous breaking process of discrete time-translational symmetry and the protection mechanism of the time crystalline state, while phase space crystals investigate the interplay of basic response modes. In section 6.1, we introduce a short history of Wilczek's time crystal based on the spontaneous breaking of continuous time-translational symmetry in a static system. As Wilczek's proposal is believed to be impossible for generic physical systems, the concept of time crystal was extended into periodically driven systems and is called *discrete* or *Floquet time crystals* nowadays. We introduce the basics of Floquet time crystals in section 6.2. In the last section 6.3, we clarify the relationship and also the difference between time crystals and phase space crystals.

6.1 Wilczek's time crystal

6.1.1 Continuous time-translation symmetry breaking

In the year 2012, Frank Wilczek proposed the idea of time crystals [1, 2]. He discussed the possibility that *continuous time-translation symmetry (CTTS)* might be spontaneously broken in a *closed time-independent* quantum mechanical system. As a consequence, some physical observable of the quantum mechanical ground state exhibits persistent periodic oscillation. Wilczek also noticed that the Heisenberg equation of motion for an operator, i.e.,

$$\langle \Psi_E | \dot{\mathcal{O}} | \Psi_E \rangle = i \langle \Psi_E | [H, \mathcal{O}] | \Psi_E \rangle = 0 \tag{6.1}$$

for any eigenstate Ψ_E of H, seems to make the possibility of quantum time crystals implausible. Nevertheless, he proposed a model consisting of N particles with contact attractive interaction on a ring threaded by an Aharonov–Bohm (AB) flux with the many-body Hamiltonian given by ($\hbar = 1$)

doi:10.1088/978-0-7503-3563-8ch6

$$H = \sum_{j=1}^{N} \frac{1}{2}\left(-i\frac{\partial}{\partial \phi_j} - \alpha\right)^2 - g_0 \sum_{j \neq k}^{N} \delta(\phi_j - \phi_k), \tag{6.2}$$

where ϕ_j is the angular coordinate of jth particle on the ring and $g_0 > 0$ is the strength of attractive contact interactions between two particles. In the spirit of mean-field approximation and taking the product ansatz of many-body wave function, i.e.,

$$\Psi(\phi_1, \dots, \phi_N) = \prod_{j=1}^{N} \psi(\phi_j), \tag{6.3}$$

the original many-body problem is reduced to the following one-body nonlinear Schrödinger equation (NLSE)

$$i\frac{\partial \psi}{\partial t} = \frac{1}{2}\left(-i\frac{\partial}{\partial \phi} - \alpha\right)^2 - g_0(N-1)|\psi|^2\psi. \tag{6.4}$$

For the case of zero flux $\alpha = 0$, it is well-known that the ground state of the above nonlinear Schrödinger equation is a bright soliton solution for $g_0(N-1) \geqslant \pi/2$ given by the hyperbolic secant function

$$\psi(\phi) \approx \mathrm{sech}\left[-\frac{g_0(N-1)}{2}(\phi - \phi_c)\right], \tag{6.5}$$

where ϕ_c is the position of the centre of mass of particles on the ring. In this case, the bright soliton breaks the continuous angular translational symmetry of the ring, but stills reserves the CTTS since the soliton solution is static (time-independent).

For the case of non-integer flux α, Wilczek was able to construct a solution of the NLSE which *rotates* with angular velocity $\omega = \alpha$ (basically taking $\phi \to \phi + \alpha t$ in the rotating frame). From the view point of Faraday's law and the 'gauge transformation', the flux-induced energy change with respect to the flux-zero ground state is $\Delta\epsilon = \alpha^2/2$ per particle. Wilczek claims this *rotating soliton* is the 'ground state' of his model. If true, we have a *time-dependent* ground state for a *time-independent* Hamiltonian, which spontaneously breaks the CTTS and satisfies Wilczek's expectation of time crystals.

However, a naive analysis indicates that the spontaneous breaking of CTTS is against the energy conservation law. For the short-range interaction, the rotating soliton generates a time-dependent force field on the ring. Imagine that we put a particle at some point on the ring when the force exerted by the soliton is weak. As the soliton is approaching the point, the particle will be accelerated and thus absorb some energy from the rotating soliton. Since the ground state has the lowest energy by definition, some positive net energy is created without lowering the energy of the rotating soliton. Wilczek's ring can be viewed as a new version of perpetual motion machine. This implies something wrong in his analysis. In the history of science, all the proposals of perpetual motion machines failed. However, several famous ones, e.g., Maxwell's demon [3] and Brownian ratchet [4], played significant roles in advancing science and still inspire new ideas today. It needs rigorous mathematical proof to disprove Wilczek's idea, which may deepen our understanding of physics laws.

6.1.2 No-go theorems

Bruno first pointed out Wilczek's rotating solution is not the true ground state for the model he proposed [5]. The correct ground state is actually a stationary state without breaking CTTS [6]. In fact, one can gauge away the flux α in NLSE (6.4) by the transformation $\psi(\phi) = e^{i\alpha\phi}\tilde{\psi}(\phi)$ with the twisted boundary condition

$$\tilde{\psi}(\phi + 2\pi) = e^{-i2\pi\alpha}\tilde{\psi}(\phi). \tag{6.6}$$

For the special case of $\alpha = 1/2$, one simply solves the NLSE of $\tilde{\psi}$ with antiperiodic boundary condition $\tilde{\psi}(\phi + 2\pi) = -\tilde{\psi}(\phi)$. The correct ground state is a stationary wave function in the form of [5, 6]

$$\tilde{\psi}(\phi) = \frac{kK}{\pi\sqrt{g_0(N-1)}}\,\mathrm{cn}(\pi^{-1}K\phi, k) \tag{6.7}$$

with total energy per particle given by

$$\epsilon = -\frac{K^2}{6\pi^2}\frac{(2k^2-1)E - (1-k^2)(3k^2-1)K}{E-(1-k^2)K}. \tag{6.8}$$

Here, $K \equiv K(k)$ and $E \equiv E(k)$ are the complete elliptic integrals of the first and second kind, and $\mathrm{cn}(u, k)$ is a Jacobi elliptic function. The elliptic modulus k is determined from the condition $[E - (1-k^2)K]K = \pi g_0(N-1)/2$. For $g_0(N-1) \gg 1$, the simple asymptotic result for the gained energy is $\Delta\epsilon = -6[g_0(N-1)]^2\exp[-\pi g_0(N-1)]$, which is negative and considerably lower than Wilczek's positive result ($\Delta\epsilon = \alpha^2/2$).

However, Bruno's simple discussion above only precludes the existence of time crystal in the soliton model but does not disprove Wilczek's general idea of time crystals for other physical systems. If the interaction between particles is not restricted to the contact form, e.g., the interaction is the long-range Coulomb interaction, one can ask whether a charge density wave (CDW) can appear and rotate as the ground state. In fact, after Wilczek's original work [7], other authors soon proposed this rotating CDW or Wigner crystal with trapped ions. However, a detailed analysis showed that the rotating CDW cannot be the ground state [8]. In the following, we introduce two 'no-go theorems' that rigorously rule out the possibility of spontaneous CTTS breaking in the ground state (or thermal equilibrium) for a broad class of systems.

6.1.2.1 Bruno's no-go theorem

In statistics, spontaneous symmetry breaking occurs in the thermodynamic limit, where the particle number goes to infinity but the particle density is fixed. A correct method to treat thermodynamic limit correctly was given by Bogoliubov [9]. One first calculates physical quantities for the system with finite size N and perturbative symmetry breaking potential v, then takes the limit $N \to \infty$ before taking the limit $v \to 0$. The wrong order of limits could lead to incorrect and unphysical results. Following Bogoliubov's method, Bruno introduced an external symmetry breaking potential rotating at angular velocity ω, and calculated the nth level energy of the system in the static frame [10]

$$E^{(\omega)}_{\alpha,\,n} = E^{(0)}_{\alpha,\,n} + \frac{1}{2}I_{\alpha,n}\omega^2 + O(\omega^3). \tag{6.9}$$

Here, α is the flux threaded through Wilczek's ring and $I_{\alpha,n}$ is the moment of inertia for the nth energy level of the system. Bruno's crucial finding is that the moment of inertia for the ground state ($n = 0$) is always positive ($I_{\alpha,n=0} > 0$), and $E^{(\omega)}_{\alpha,\,n=0} > E^{(0)}_{\alpha,\,n=0}$ for any finite value of $\omega \neq 0$. For the system in the thermal equilibrium at inverse temperature β, the free energy has a similar inequality, i.e., $\mathcal{F}^{(\omega)}_{\phi,\,\beta} > \mathcal{F}^{(0)}_{\phi,\,\beta}$ for $\omega \neq 0$. The above two inequalities strictly prohibit the existence of spontaneously rotating time crystals for the ground state and thermal equilibrium state, which is the 'no-go theorem' proved by Bruno. Note that the moment of inertia for an excited state $I_{\phi,n>0}$ may be negative. This allows for the existence of time crystal at some excited eigenstate [11, 12] or time-crystal Sisyphus dynamics near the energy minimum [13].

6.1.2.2 WO's no-go theorem

Since Bruno's proof is restricted to a special model of Aharonov–Bohm ring, Wilczek and others later proposed new types of time crystals to avoid Bruno's no-go theorem [14, 15]. A more general no-go theorem was given by Watanabe and Oshikawa (WO) [16]. WO first presented a precise definition for the time crystals using the local order parameter $\phi(\vec{x}, t)$. A system with Hamiltonian H is in the time crystal state if the correlation function given by

$$\lim_{V\to\infty}\langle e^{iHt}\phi(\vec{x}, 0)e^{-iHt}\phi(0, 0)\rangle \to f(t) \tag{6.10}$$

exhibits a nontrivial periodic oscillation in time t for large $|\vec{x}|$ (much greater than any microscopic scales). In terms of the integrated order parameter $\Phi \equiv \int_V d^dx\phi(\vec{x}, 0)$, the correlation function reads equivalently $\lim_{V\to\infty}\langle e^{iHt}\Phi e^{-iHt}\Phi\rangle / V^2 \to f(t)$. At zero temperature, WO proved the following equality for the ground state

$$\frac{1}{V^2}|\langle 0|\Phi e^{-i(H-E_0)t}\Phi|0\rangle - \langle 0|\Phi^2|0\rangle| \leqslant C\frac{t}{V}, \tag{6.11}$$

where E_0 is the ground state energy, and constant C depends on Φ and H but not on t and V. The above inequality holds for the interaction that decays exponentially as a function of the distance or decays as power-law $r^{-\alpha}(\alpha > 0)$. Therefore, for any fixed time t, the correlation function $f(t)$ remains constant in the limit $V \to \infty$. The conclusion can be extended to the case of canonical ensemble at finite temperature by the help of Lieb–Robinson bound [17]. A related definition of time translation symmetry breaking (TTSB) was discussed in the language of C^* algebras [18], which recovers WO's result for the absence of TTSB in equilibrium systems.

It is still an open question of if one can bypass the limitations imposed by WO's no-go theorem and reach the desirable phase of a quantum time crystal in the ground state of an isolated quantum system without a driver [19].

6.2 Floquet time crystals

6.2.1 Discrete time-translation symmetry breaking

The no-go theorems discussed above do not exclude the possibility of spontaneous time-translational symmetry breaking in quantum systems far away from equilibrium. For the intrinsically nonequilibrium setting of a periodically driven (Floquet) system, the Hamiltonian has the *discrete time-translation symmetry (DTTS)* such that $H(t) = H(t + T_d)$ with $T_d = 2\pi/\omega_d$ the period of driving field. It was proposed that the DTTS can be spontaneously broken in periodically driven atomic systems [20] and spin chains [21–24].

To clarify spontaneous DTTS breaking, we define a *short-ranged correlated* state $|\psi\rangle$ if, for any local operator $\phi(\vec{x})$, the state satisfies [22]

$$\langle\psi|\phi(\vec{x})\phi(\vec{x}\,')|\psi\rangle \rightarrow \langle\psi|\phi(\vec{x})|\psi\rangle\langle\psi|\phi(\vec{x}\,')|\psi\rangle \tag{6.12}$$

as $|\vec{x} - \vec{x}\,'| \rightarrow \infty$. The DTTS is said to be broken if for every 'short-range correlated' state $|\psi\rangle$, there exists an operator Φ and integer $n > 1$ such that

$$\langle\psi(nT_d)|\Phi|\psi(nT_d)\rangle = \langle\psi(0)|\Phi|\psi(0)\rangle, \tag{6.13}$$

namely, the system responds at a fraction ω_d/n $(n > 1)$ of the original driving frequency [22]. We call the robust short-ranged correlated state with broken DTTS the *Floquet time crystal* (FTC) or *discrete time crystal* (DTC).

Floquet time crystals have been realised in various experimental platforms, e.g., trapped ions [25], diamond nitrogen vacancy centres [26], nuclear spins [27–29], superfluid ^{3}He–B (magnon BEC) [12] and ultracold atoms [30]. For more recent progress on the study of time crystals, we refer the readers to other review papers [31–33] and the references therein.

6.2.2 Subharmonic modes

In order to better understand the formation of Floquet time crystals, we consider a concrete example of one dimensional chain of N spin-half particles. The spin chain is subjected to a binary driving protocol with the Hamiltonian given by

$$H(t) = \begin{cases} H_1 \equiv (1 - \delta)B^x \sum_i \sigma_i^x, & 0 \leqslant t < \tau\hbar/B^x \\ H_2 \equiv \sum_i J\vec{\sigma}_i \cdot \vec{\sigma}_{i+1} + B_i^z\sigma_i^z, & \tau\hbar/B^x \leqslant t < T_d. \end{cases} \tag{6.14}$$

Here, $\vec{\sigma}_i \equiv (\sigma_i^x, \sigma_i^y, \sigma_i^z)$ represents Pauli operators, $B_i^z \in [0, W]$ is a random longitudinal field, δ is the imperfection of the uniform transversal field B^x, and $T_d > \tau\hbar/B^x$ is the driving period with $\tau = \pi/n$ with $n \in \mathbb{N}$ the dimensionless time interval.

For the case of perfect driving (imperfection parameter $\delta = 0$) and clean system (disorder strength $W = 0$), the Floquet unitary defined as the time evolution during one time period, see equation (2.130), has the form ($\hbar = 1$)

$$U_f = e^{-i(T_d - \tau)\sum_i J\vec{\sigma}_i \cdot \vec{\sigma}_{i+1}} R_\tau \quad \text{with} \quad R_\tau \equiv e^{-i\tau \sum_i \sigma_i^x}. \tag{6.15}$$

Here, we have chosen the Floquet period $T = T_d$. Since the operator R_τ rotates all the spins, the short-range correlated state $|\phi_0\rangle \equiv |\uparrow\uparrow\uparrow\cdots\uparrow\rangle$ is NOT the eigenstate of U_f. However, the real Floquet eigenstates $|\psi_m\rangle$ $(m = 0, 1, \ldots, n-1)$ can be constructed from ϕ_0 in the form

$$|\psi_m\rangle \propto \sum_{q=0}^{n-1} e^{i2mq\tau} R_\tau^q |\phi_0\rangle \equiv \sum_{q=0}^{n-1} e^{i2mq\tau} |\phi_q\rangle \tag{6.16}$$

with short-range correlated states

$$|\phi_q\rangle \equiv R_\tau^q |\uparrow\uparrow\uparrow\cdots\uparrow\rangle \ \ (q = 0, 1, \ldots, n-1). \tag{6.17}$$

From $R_\tau|\psi_m\rangle = e^{-i2m\tau}|\psi_m\rangle$, all the Floquet eigenstates given by equation (6.16) have $\mathbb{Z}_n$ symmetry, i.e., $R_\tau^n|\psi_m\rangle = |\psi_m\rangle$. Note that the interaction part in H_2 is invariant by the unitary transformation of operator R_τ. The quasienergy of Floquet eigenstate $|\psi_m\rangle$ can be determined from $U_f|\psi_m\rangle = e^{-i\epsilon_m T_d}|\psi_m\rangle$, i.e.,

$$\epsilon_m = NJ\left(1 - \frac{\omega_d}{2n}\right) + m\frac{\omega_d}{n}, \tag{6.18}$$

where $\omega_d = 2\pi/T_d$ is the driving frequency. Therefore, the set of n Floquet eigenstates $|\psi_m\rangle$ $(m = 0, 1, \ldots, n-1)$ form a tuple in the quasienergy spectrum with equidistant quantised quasienergy ω_d/n as illustrated in figure 6.1(left). Note that the short-ranged correlated states $|\phi_q\rangle$ given by equation (6.17) are not orthogonal for a finite-size system, i.e.,

$$\langle\phi_q|\phi_{q'}\rangle = \left[\cos\frac{(q - q')\pi}{n}\right]^N \neq 0 \quad \text{for} \quad q \neq q'.$$

However, in the thermodynamic limit $N \to \infty$, all the short-ranged correlated states $|\phi_q\rangle$ are orthogonal $\langle\phi_q|\phi_{q'}\rangle = \lim_{N\to\infty}\left[\cos\frac{(q - q')\pi}{n}\right]^N \to \delta_{qq'}$.

The Floquet eigenstates $|\psi_m\rangle$ given by equation (6.16) are long-range correlated states and are unlikely to survive in the thermodynamic limit (similar to multi-component Schrödinger-cat states). If the system is initially prepared in one Floquet eigenstate $|\psi_m\rangle$, the system will collapse randomly into one of the short-range correlated basis $|\phi_q\rangle$. As consequence, the $\mathbb{Z}_n$ symmetry of $|\psi_m\rangle$ is spontaneously broken in the thermodynamic limit. From equation (6.16), we can express the short-range correlated states with Floquet eigenstates

$$|\phi_q\rangle = \frac{1}{\sqrt{n}} \sum_{m=0}^{n-1} e^{-i2mq\tau}|\psi_m\rangle, \ \text{with} \quad q = 0, 1, \ldots, n-1. \tag{6.19}$$

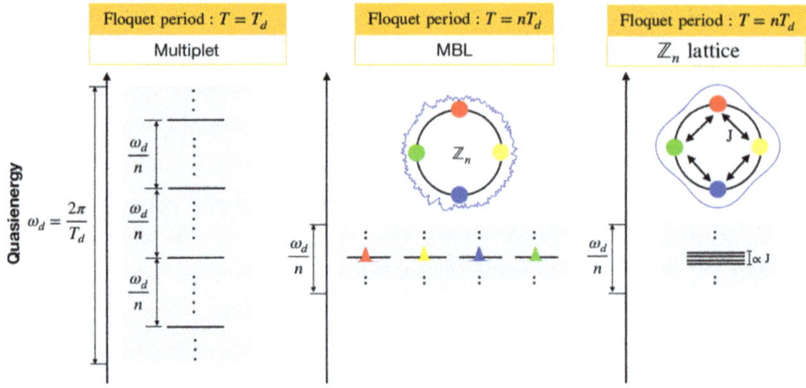

Figure 6.1. Floquet time crystals. (Left) An n-component multiplet of Floquet eigenstates emerges in the quasienergy spectrum with the choice of Floquet period equal to the driving period $T = T_d$. (Middle) The n Floquet multiplets become degenerate (i.e., $\mathbb{Z}_n$ symmetry) with the choice of Floquet period $T = nT_d$. In the thermodynamic limit, the $\mathbb{Z}_n$ symmetry of Floquet eigenstates are spontaneously broken into subharmonic modes (coloured wave packets), which can be stabilised by many-body localisation (MBL) with disorders (blue noisy curve). (Right) If there is an additional $\mathbb{Z}_n$-lattice potential (blue curve), the subharmonic modes tunnel to its nearest-neighbour with rate J resulting in a Floquet band with width $\propto J$. In the limit of $J \to 0$, the Floquet eigenstates are broken into robust subharmonic modes localised on the $\mathbb{Z}_n$-lattice sites. Reprinted from [34].

Using equation (6.18), we have the evolution of $|\phi_q(t)\rangle$ at stroboscopic time steps $t = pT_d$

$$|\phi_q(pT_d)\rangle = \frac{1}{\sqrt{n}} e^{-iNJ(1-\frac{\omega_d}{2n})pT_d} \sum_{m=0}^{n-1} e^{-i2mq\frac{\pi}{n} - im\frac{\omega_d}{n}pT_d} |\psi_m(0)\rangle. \tag{6.20}$$

We see that $\langle\phi_q(nT_d)|\hat{\Phi}|\phi_q(nT_d)\rangle = \langle\phi_q(0)|\hat{\Phi}|\phi_q(0)\rangle$ with integer $n > 1$ satisfying the condition equation (6.13). Therefore, in the thermodynamic limit $N \to +\infty$, the DTTS is spontaneously broken accompanied by the breaking of a global internal $\mathbb{Z}_n$ symmetry described by R_τ. The symmetry-broken short-range correlated states are also called *subharmonic modes* since their oscillating frequency is a fraction ω_d/n of the driving frequency ω_d.

Since the original time-dependent Hamiltonian also satisfies $H(t) = H(t + nT_d)$, we have the freedom to choose the Floquet period $T = nT_d$. In this setting, the Floquet Hamiltonian introduced by equation (2.131) can be formally rewritten as

$$H_F = i\frac{1}{nT_d} \ln\left(\mathcal{T}\exp\left[-i\int_0^{nT_d} H(t)dt\right]\right). \tag{6.21}$$

Under this choice, the n Floquet multiplets are 'degenerate' within the Floquet–Brillouin zone $0 \leqslant \epsilon_\alpha < \omega_d/n$ as illustrated by the black lines in figure 6.1(middle). They are long-range correlated eigenstates analogous to multi-component Schrödinger-cat states with $\mathbb{Z}_n$ symmetry as shown by equation (6.16). In the thermodynamic limit, the n long-range correlated states randomly collapse into one of the n short-range correlated states, which are illustrated by the coloured wave packets in figure 6.1(middle). As a consequence, the $\mathbb{Z}_n$ symmetry of long-range correlated eigenstates is spontaneously broken, and the resultant n short-range correlated states exhibit oscillations with the same time period nT but with phases differed by $2\pi/n$, cf. equations (6.19) and (6.20). We arrange the n short-range correlated states on a $\mathbb{Z}_n$ circle as shown in figure 6.1(middle). The emergence of a multiplet of Floquet eigenstates with equidistant quasienergy ω_d/n as illustrated in figure 6.1(left) is the prerequisite for DTTS breaking [21–23].

Subharmonic modes in quantum gas Besides spin model, spontaneous breaking of DTTS and the resultant subharmonic modes can also occur in the driven system of quantum gas. In [20], K Sacha considered N atoms with contact interaction $g_0\delta(x_i - x_j)$ and subject to a Floquet Hamiltonian $H_0(t) = H_0(t + 2\pi/\omega_d)$. In the spirit of mean-field approximation, all the atoms occupy the same single-particle wave function. The dynamics of the mean-field state is described by the Gross–Pitaevskii equation (GPE)

$$\left[H_0(t) + g_0 N|\psi(x, t)|^2 - i\frac{\partial}{\partial t}\right]\psi(x, t) = \mu\psi(x, t). \tag{6.22}$$

Consider the special case of $\mathbb{Z}_2$ lattice symmetry. For $g_0 N \to 0$, the eigensolutions of GPE are identical to the corresponding Floquet states of the linear system, i.e., $\psi_1(x, t)$ and $\psi_2(x, t)$ with quasienergies ϵ_1 and $\epsilon_2 = \epsilon_1 + \omega_d/2 + |J|$, where $|J|$ is the tunnelling amplitude. One can define the two local Wannier states by

$$\phi_{1,2}(x, t) = [\psi_1(x, t) \pm e^{-i\omega_d t/2}\,\psi_2(x, t)]/\sqrt{2}.$$

If the interaction is weak such that $N|U - 2U_{12}| < |J|$ where $U = g_0 \int dxdt|\phi_1(x, t)|^4$ and $U_{12} = g_0 \int dxdt|\phi_1(x, t)|^2|\phi_2(x, t)|^2$, GPE has two stable solutions:

$$\psi_\pm = (\phi_1 \pm \phi_2)/\sqrt{2}.$$

If the interaction is strong such that $N|U - 2U_{12}| > |J|$ and $g_0 < 0$ (attractive interaction), solution ψ_+ becomes unstable and two new stable solutions are

$$\psi_{\pm v}(x, t) = \sqrt{\frac{1 \pm v}{2}}\,\phi_1(x, t) + \sqrt{\frac{1 \mp v}{2}}\,\phi_2(x, t)$$

with $v = \sqrt{1 - |J|^2/|N(U - 2U_{12})|^2}$. The solutions ψ_v and ψ_{-v} have the same extremal value (degenerate) and can be measured in an experiment with equal probability (which of them is realised is determined in a spontaneous symmetry-breaking process of DTTS). As the tunnelling amplitude $|J|$ is usually very small, even very weak particle interactions lead to breaking of DTTS in the mean-field description. The $\mathbb{Z}_n$ symmetry breaking (period-n tuppling) was discussed, e.g., in the ultracold atoms bouncing on an oscillating mirror [35], the n-hands clock model [36, 37], ultracold bosons in a ring lattice [38] and long-range interacting systems [39]. The fractional time crystals, whose period of motion is a rational fraction of the driving period, was also proposed and studied in [40].

6.2.3 Stabilising mechanisms

In fact, the phenomena of subharmonic response or the period multiplication in classical systems have been unknown for quite a long time. The history is dated back to the study of Faraday, i.e., the famous Faraday waves [41]. The appearance of subharmonic modes in classical systems is the consequence of nonlinear dynamics. However, because the Schrödinger equation describing quantum dynamics is *linear* at the fundamental level, the appearance of subharmonic response in quantum systems needs renewed explanation. This is the spontaneous DTTS breaking discussed above. More seriously, as there is no nonlinearity protection as in classical dynamics, the subharmonic modes of a generic quantum many-body system are generally unstable since the system is heated up to a featureless infinite-temperature state by the driving [42–44]. As an eligible Floquet time crystal, it needs some protection mechanism to stabilise the subharmonic mode. In the following, we introduce some typical stabilisation mechanisms in the current study of time crystals.

6.2.3.1 Many-body localisation

In a closed quantum many-body system, disorder (e.g., setting $W > 0$ for random field B_i^z in equation (6.14)) can be introduced to stabilise the subharmonic modes [21–24] via the mechanism of many-body localisation (MBL) [45–48]. To distinguish the MBL phase from the thermalised state, one can calculate the quasienergy statistics ratio given by

$$\langle r \rangle = \frac{\min(\Delta_m, \Delta_{m+1})}{\max(\Delta_m, \Delta_{m+1})}, \tag{6.23}$$

where $\Delta_m = \epsilon_{m+1} - \epsilon_m$ is the mth quasienergy gap [21, 49], by averaging over both disorder and the quasienergy spectrum. For a fixed disorder W, if the interaction is larger than a critical value $J > J_c$, the system is in the thermal phase with the energy level statistics ratio approaching the circular orthogonal ensemble of $\langle r \rangle \approx 0.527$. When $J < J_c$, the disorder is dominant and the system is in the MBL phase with the energy level statistics ratio approaching the Poisson limit of $\langle r \rangle \approx 0.386$ [43]. The whole MBL phase can be further divided into two subregions depending on the driving imperfection δ [24]. If the imperfection is larger than the critical point $\delta > \delta_c$, the MBL state responses at the driving frequency and the DTTS is not broken. If the driving imperfection is not too strong, i.e., for a certain range $\delta \in (0, \delta_c)$, the DTTS is broken and the resultant MBL state is an eligible Floquet time crystal. Figure 6.2 shows the schematic phase diagram of Floquet time crystal as functions of interaction strength J/W and driving imperfections δ.

6.2.3.2 Prethermalisation

Although a generic driven many-body clean system (without disorder) is eventually heated up a trivial featureless infinite-temperature state with no long-range order, the heating process may take an exponentially long time provided that the driving frequency is much larger than the local energy scales in the system [50–53]. The heating is a high-order process as the system needs many local rearrangements of the degrees of freedom to absorb an 'energy quanta' from the driving field. The transient state with extremely long life time is called a prethermal state, and the DTTS broken state of Floquet time crystals in this long time prethermal process is called *prethermal time crystal* [27, 28, 39, 54–59].

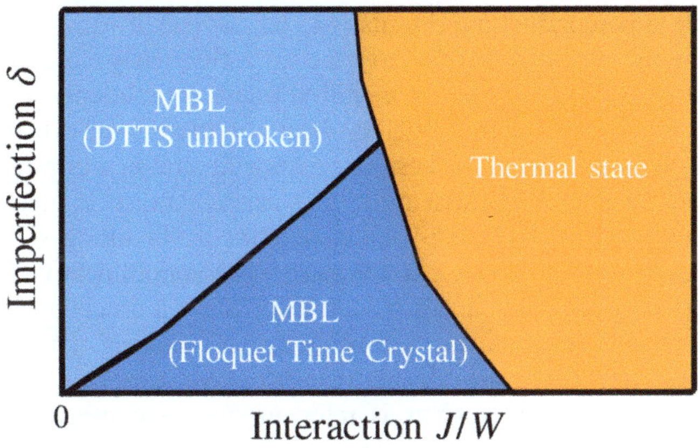

Figure 6.2. Schematic phase diagram of Floquet time crystal. Depending on the interaction strength J/W, the system can be a thermal state or a many-body localised (MBL) state. The MBL phase is further divided into the regular phase (DTTS is preserved) and the time crystal phase (DTTS is broken) depending on the driving imperfections δ. Produced schematically according to the results in [24].

6.2.3.3 *Dissipation*

In fact, the spontaneous breaking of DTTS has been well-known for quite a long time as the self-organised phenomena in classical driven-dissipative systems, e.g., synchronisation in Kuramoto model [60] and Belousov–Zhabotinsky reaction [61]. We emphasise that the Floquet time crystals by the original definition are the quantum many-body phenomena emerging in *closed* systems not in open systems. Nevertheless, one can lose the constraint of time crystals and extend the study of classical self-organised phenomena into quantum driven-dissipative systems. Actually, the spontaneous breaking of DTTS was already discussed even before Wilczek's seminal work in a modulated dissipative cold atom system [62], and the DTTS symmetry breaking has been reported in the experiment [63]. The stable time crystalline state with broken DTTS by coupling the Floquet many-body system to a cold bath is called *dissipative time crystals* [58].

6.2.3.4 *Nonlinearity*

Although Schrödinger's equation is linear at the fundamental level, some effective nonlinear effect after DTTS breaking can stabilise the subharmonic modes. Imagine the Floquet Hamiltonian (6.21) has a $\mathbb{Z}_n$-lattice potential along the $\mathbb{Z}_n$ circle as illustrated by the blue curve in figure 6.1(right). In quantum mechanics, the Floquet eigenstates are the superposition of localised states near the lattice sites (Wannier functions), which can tunnel to their neighbouring sites with hopping rate J. As discussed in section 2.2.2, the system can be described by a tight-binding model with Floquet bandwidth $\propto |J|$. In the classical limit $\lambda \to 0$ and thus $J \to 0$, the Floquet eigenstates become degenerate and form a multiplet ready for symmetry breaking. At this point, any nonequilibrium fluctuations can induce the Floquet eigenstates to collapse into one of the localised states with equal probability. The broken state is protected by the nonlinearity of local potential well at the $\mathbb{Z}_n$ lattice sites and oscillates with time period of nT forming Floquet time crystal. This period-multiplication phenomena has been observed in the experiment with superconducting circuits [64, 65], where n robust output radiation modes (subharmonic oscillations [66]) have the same frequency close to a fraction ω_d/n of the driving frequency but evenly $2\pi/n$-shifted phase components. The $\mathbb{Z}_2$ symmetry breaking of parametrically driven model has been reported in the experiment with cold atoms [63] and analysed in [62]. These experimental findings display classical fixed points and the hopping between them induced by classical noise. In quantum mechanics where the tunnelling rate is finite $|J| > 0$, the spontaneous breaking of DTTS can occur by the attractive interaction between atoms as shown the nonlinear GPE (6.22). In this scenario, the $|\psi|^2$-dependent nonlinear part in GPE plays a crucial role to stabilise the Floquet time crystal. The nonlinearity mechanism of realising time crystals has also been studied in the driven Lipkin–Meshkov–Glick (LMG) model [54], where only the collective degree of freedom of spins is relevant due to conservation laws. In the thermodynamic limit, the collective spins behave classically and the subharmonic response is protected by nonlinear resonances.

Looking for a new protection mechanism of DTTS breaking is still an active research in the study of time crystals, e.g., Floquet dynamical symmetry (FDS) [67].

6.2.4 Time crystals in multi-dimensional systems

We have discussed the time crystals in 1D systems, where the DTTS manifests itself as the Z_n symmetry of subharmonic modes. Similar discussion can be readily extended into higher-dimensional systems [68]. Let us first consider the motion of a particle in 3D described by a separable time-independent Hamiltonian

$$H_0 = H_0^{(1)}(q_1, p_1) + H_0^{(2)}(q_2, p_2) + H_0^{(3)}(q_3, p_3), \tag{6.24}$$

where the three independent Hamiltonians on the right-hand side are all integrable (see section 2.2.1). Following the procedure in section 3.1.1, we introduce a tuple of action variables $\mathbf{I} = (I_1, I_2, I_3)$ and a tuple of conjugate angle variables $\boldsymbol{\theta} = (\theta_1, \theta_2, \theta_3)$ to construct a six dimensional phase space. The action variables are defined by the following integrals along the isoenergetic contours for the three independent degrees of freedom

$$I_i \equiv (2\pi)^{-1} \oint p_i \, \mathrm{d}q_i, \qquad i = 1, 2, 3.$$

The angle variables are given via the action functions, cf. equation (2.67). In this way, we rewrite the Hamiltonian as a function of action $H_0(\mathbf{I})$ that is independent of the angle variables. According to the canonical equation of motion (2.14), we have

$$\theta_i(t) = \Omega_i t + \theta_i(0),$$

where $\Omega_i = \partial H_0(\mathbf{I})/\partial I_i$ ($i = 1, 2, 3$) represents the frequencies of the regular motion orbit along three independent directions. We then introduce a time-dependent perturbative Hamiltonian in the form of

$$H_\epsilon = \epsilon_1 h_1(q_1) f_1(t) + \epsilon_2 h_2(q_2) f_2(t) + \epsilon_3 h_3(q_3) f_3(t), \tag{6.25}$$

where the time periodic function $f_j(t)$ ($j = 1, 2, 3$) is real and reads $f_j(t + 2\pi/\omega_d) = f_j(t) = \sum_k f_k^{(j)} e^{ik\omega_d t}$ with $f_k^{(j)} = f_{-k}^{(j)*}$. The spatial function h_j

($i = 1, 2, 3$) can be expressed with action and angle variables by $h_j = \sum_l h_l^{(j)}(I_j)e^{il\theta_j}$. The whole single-particle Hamiltonian is then given by

$$H_s(t) = H_0(I_1, I_2, I_3) + \sum_{j,l,k} \epsilon_j h_l^{(j)}(I_j) f_k^{(j)} e^{il\theta_j + ik\omega_d t}. \tag{6.26}$$

If we set the initial condition (I_1, I_2, I_3) such that the driving frequency resonates with all the three rational orbits, i.e.,

$$\omega_d = n_1\Omega_1 = n_2\Omega_2 = n_3\Omega_3, \quad n_, n_2, n_3 \in \mathbb{N},$$

the motion of the particle in the vicinity of the resonant orbit can be separated into the fast oscillating and slow varying parts. The slow motion can be described by a new set of angle variables $\mathbf{\Theta} = (\Theta_1, \Theta_2, \Theta_3)$ with

$$\Theta_i = \theta_i(t) - \Omega_i t, \quad i = 1, 2, 3. \tag{6.27}$$

Using the generating function of the second kind $G_2 = \sum_{i=1,2,3} I_i(\theta_i - \Omega_i t)$, cf. section 2.1.3, we transform into the rotating frame and obtain the effective time-independent (RWA) Hamiltonian from equation (6.26)

$$\begin{aligned} H_{\mathrm{eff}} &= H_0(\mathbf{I}) - \Omega_1 I_1 - \Omega_2 I_2 - \Omega_3 I_3 + V_{\mathrm{eff}}(\mathbf{I}, \Theta_1, \Theta_2, \Theta_3) \\ &\approx \sum_i \frac{1}{2m_{\mathrm{eff}}^{(i)}} P_{\Theta_i}^2 + V_{\mathrm{eff}}(\Theta_1, \Theta_2, \Theta_3). \end{aligned} \tag{6.28}$$

Here, the potential $V_{\mathrm{eff}}(\mathbf{I}, \mathbf{\Theta}) \equiv \sum_{j,k} \epsilon_j h_{kn_j}^{(j)}(I_j) f_{-k}^{(j)} e^{ikn_j\Theta_j}$ is a periodic potential in the three dimensional space $(\Theta_1, \Theta_2, \Theta_3)$ in the rotating frame. This effective Hamiltonian essentially describes a particle moving in 3D space with periodic boundary conditions (3D torus). The time-independent effective lattice potential $V_{\mathrm{eff}}(\mathbf{I}, \mathbf{\Theta})$ can be engineered by tuning the Fourier components $f_k^{(j)}$ and $h_l^{(j)}$ in the experiments. In the second line of equation (6.28), we have expanded $H_{\mathrm{eff}}(\mathbf{I}, \mathbf{\Theta})$ near the resonant orbits (I_1, I_2, I_3) and defined the local momentum $P_{\Theta_i} = \Delta I_i$, cf. equation (3.29) for the 1D case. The effective mass is given by $m_{\mathrm{eff}}^{(i)} = \partial^2 H_{\mathrm{eff}}/\partial I_i^2$. Note that, in [68], the authors used the perturbation which is the product (not the sum) of the terms in equation (6.25) to realise the 3D Anderson localized–delocalized transition.

6.2.4.1 2D gravitational bouncer

In section 3.1.1, we introduced the model of 1D gravitational bouncer. We now discuss an extended model of 2D gravitational bouncer, that is, atoms bouncing between two mirrors that oscillate harmonically in two orthogonal directions [70, 71]. The experimental setup is sketched in figure 6.3(a), where two orthogonal mirrors are inclined at an angle of $\pi/4$ to the gravitational force. We set the direction along the two mirrors as x and y axes respectively, and the gravitational force $\mathbf{F}_g$ points in the $-(\mathbf{x} + \mathbf{y})$ direction. The single-atom Hamiltonian is given by [71, 72]

$$H_s = \frac{1}{2}p_x^2 + \frac{1}{2}p_y^2 + x + y + \epsilon_x x \cos(\omega_d t + \Delta\phi) + \epsilon_y y \cos(\omega_d t), \tag{6.29}$$

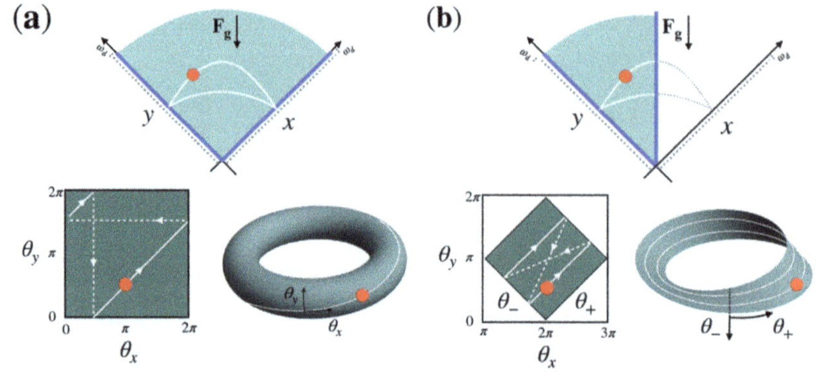

Figure 6.3. Schematic plot of 2D gravitational bouncer. An atom (red dot) bounces between two oscillating mirrors (thick blue lines) with gravitational force $\mathbf{F}_g$. (a) Orthogonal mirrors. The white line in the upper panel represents a typical trajectory for static mirrors. In the lower left panel of (θ_x, θ_y) plane, we show the corresponding trajectory (solid white lines) and the periodic boundary conditions for the angle variable $\theta_{x(y)}$ (dashed white lines). The motion is identical to the closed periodic orbit on the 2D (θ_x, θ_y) torus in the lower right panel. (b) Two mirrors with a $\pi/4$ wedge. The typical trajectory is folded by the vertical mirror. In the lower left panel, the corresponding trajectory (solid white lines) is cut by the vertical mirror and connected via the white dashed lines. This constraint defines a Möbius strip geometry as shown in the lower right panel [69]. Produced following the models and methods in [69].

where $\Delta\phi$ is the relative phase, and $\epsilon_{x,y}$ are the amplitudes of two mirrors' oscillations. Assuming that the resonance condition $\omega_d = s_x\Omega_x = s_y\Omega_y$ is fulfilled, we obtain the effective Hamiltonian from equation (6.28)

$$H_{\text{eff}} \approx \frac{1}{2m_{\text{eff}}^{(x)}}P_x^2 + \frac{1}{2m_{\text{eff}}^{(y)}}P_y^2 + V_0[\cos(s_x\Theta_x) + \cos(s_y\Theta_y)], \tag{6.30}$$

which describes a particle in a two dimensional square lattice. Due to the discrete translational symmetry, the Bloch eigenstates of H_{eff} are given by

$$\psi_{k_x,k_y}(\Theta_x, \Theta_y) = e^{i(k_x\Theta_x + k_y\Theta_y)}u_{k_x}(\Theta_x)u_{k_y}(\Theta_y)$$

with the periodic functions $u_\alpha(\Theta_\alpha) = u_\alpha(\Theta_\alpha + 2\pi/s_\alpha)$ $(\alpha = x, y)$, and the quantum numbers $k_\alpha = 2\pi j_\alpha/s_\alpha$ $(j_\alpha = 0, 1, \ldots, s_\alpha - 1)$. In the laboratory frame, the Bloch waves are time-dependent and can be obtained from equation (6.27), i.e.,

$$\psi_{k_x,k_y}(\theta_x, \theta_y, t) = e^{i(k_x\theta_x + k_y\theta_y) - i(k_x/s_x + k_y/s_y)\omega_d t}u_{k_x}(\theta_x - \omega_d t/s_x)u_{k_y}(\theta_y - \omega_d t/s_y). \tag{6.31}$$

For a fixed position in lab frame, say the position at (θ_x, θ_y), we will observe the periodic change of the amplitude square of Bloch functions (the probability to observe one atom) with exactly the driving period, i.e., $|\Psi_{k_x,k_y}(\theta_x, \theta_y, t)|^2 = |\Psi_{k_x,k_y}(\theta_x, \theta_y, t + 2\pi/\omega_d)|^2$, without breaking DTTS. If we restrict ourselves to the lowest energy band, we can define the Wannier states at the local minimum sites $\mathbf{j} = (j_x, j_y)$ of $V_{\text{eff}}(\Theta_x, \Theta_y)$, i.e.,

$$W_{\mathbf{j}}(\Theta_x, \Theta_y) = w_x(\Theta_x - 2\pi j_x/s_x)w_y(\Theta_y - 2\pi j_y/s_y). \tag{6.32}$$

In the laboratory frame the Wannier states describe localised wave packets moving along resonant trajectories. There are in total $s = s_x s_y$ sites of the effective potential in the moving frame corresponding to $s = s_x s_y$ Wannier wave packets. After each driving period $2\pi/\omega_d$, a single Wannier wave packet moves into another lattice site, and thus has longer time period $T = s_x s_y 2\pi/\omega_d$ breaking DTTS.

Next, we consider the many-body case, i.e., N bosons are bouncing between the mirrors and interact via attractive contact interaction $g_0\delta(\mathbf{r}_i - \mathbf{r}_j)$ ($g_0 < 0$). Following the procedure introduced in section 4.2.4 for the 1D case, we obtain the Bose–Hubbard Hamiltonian for the 2D case [71]

$$H_{\mathrm{BH}} \approx -\frac{1}{2}\sum_{\langle \mathbf{i},\mathbf{j}\rangle} J_{\mathbf{ij}} a_{\mathbf{i}}^{\dagger} a_{\mathbf{j}} + \frac{1}{2}\sum_{\mathbf{i},\mathbf{j}} U_{\mathbf{ij}} a_{\mathbf{i}}^{\dagger} a_{\mathbf{j}}^{\dagger} a_{\mathbf{j}} a_{\mathbf{i}}. \tag{6.33}$$

Here $a_{\mathbf{j}}$ are the standard bosonic annihilation operators of Wannier states at lattice site $\mathbf{j} = (j_x, j_y)$. The nearest-neighbour tunnelling rate is given by [71]

$$J_{\mathbf{ij}} = -\frac{2}{T}\int_0^T dt \langle W_{\mathbf{i}}|(H_s - i\partial_t)|W_{\mathbf{j}}\rangle, \tag{6.34}$$

and the coefficients of the off-site effective interactions ($\mathbf{i} \neq \mathbf{j}$) are given by

$$U_{\mathbf{ij}} = \frac{2}{T} g_0 \int_0^T dt \int |W_{\mathbf{i}}(t)|^2 |W_{\mathbf{j}}(t)|^2. \tag{6.35}$$

Here, the spatial integral is performed in the rest (θ_x, θ_y) or (x, y) frame, cf. equation (4.48) in section 4.2.2 for 1D case. For on-site case ($\mathbf{i} = \mathbf{j}$), the effective interaction is [70]

$$U_{\mathbf{ii}} = \frac{1}{T} g_0 \int_0^T dt \int |W_{\mathbf{i}}(t)|^4,$$

which is typically at least an order of magnitude larger than the off-site interaction $U_{\mathbf{ij}}$ ($\mathbf{i} \neq \mathbf{j}$).

For negligible interactions between atoms, the ground state of the Bose–Hubbard Hamiltonian (6.33) is a Bose–Einstein condensate (in the rest frame)

$$\Psi_G(\boldsymbol{\theta}_1, \dots, \boldsymbol{\theta}_N, t) = \prod_{j=1}^{N} \psi_0(\boldsymbol{\theta}_j, t) \tag{6.36}$$

where $\boldsymbol{\theta}_i = (\theta_x, \theta_y)$ is the position of atom $\mathbf{i}$ in the rest (θ_x, θ_y) space. This means all the atoms occupy the lowest-energy Bloch state $\psi_0 \propto \sum_{\mathbf{i}} W_{\mathbf{i}}(\boldsymbol{\theta}_i, t)$. Although each Wannier state evolves with the time period $s_x s_y 2\pi/\omega_d$ in the lab frame, the ground state $\Psi_G(\boldsymbol{\theta}_1, \dots, \boldsymbol{\theta}_N, t)$ has shorter time period $2\pi/\omega_d$ since the Wannier states exchange their positions after each driving period.

For the finite attractive interaction, the many-body dynamics in the mean-field approximation can be described by a nonlinear Schrödinger equation, cf. equation (6.22), in 2D space. When the attractive interactions between atoms are sufficiently strong, it is energetically favourable to group all atoms in a single localised wave packet $W_{\mathbf{i}}(\boldsymbol{\theta}, t)$. Then, the preferred many-body state is

$$\tilde{\Psi}(\boldsymbol{\theta}_1, \ldots, \boldsymbol{\theta}_N, t) = \prod_{j=1}^{N} W_{\mathbf{i}}(\boldsymbol{\theta}_j, t) \tag{6.37}$$

where **i** is arbitrary. According to the Floquet theory introduced in section 2.3.1, the DTTS of the Hamiltonian requires that all Floquet eigenstates must evolve with the period of the driving $2\pi/\omega_d$. As a result, the state in the form of (6.37) cannot be the eigenstate of the system. The true ground state that reserves DTTS must take the form $\Psi_G \propto \sum_{\mathbf{i}}\prod_{j=1}^{N} W_{\mathbf{i}}(\boldsymbol{\theta}_j, t)$, which is a macroscopic superposition of Bose–Einstein condensates. However, such a macroscopic superposition is extremely fragile and an arbitrary perturbation, e.g., measuring the position of one atom, makes the ground state collapse to one of the Bose–Einstein condensates

$$\Psi_G \to \tilde{\Psi} = \prod_{j=1}^{N} W_{\mathbf{i}}(\boldsymbol{\theta}_j, t), \tag{6.38}$$

which is the subharmonic mode evolving with the period $T = s_x s_y 2\pi/\omega_d$ and thus breaks DTTS of the many-body Hamiltonian. This is the process of spontaneous breaking of DTTS in 2D quantum many-body system. The DTTS broken subharmonic state (6.38) fulfils the requirement of time crystal as it is robust due to the effective nonlinearity of the periodic lattice potential $V_{\text{eff}}(\Theta_x, \Theta_y)$ in the Hamiltonian (6.30).

In a recent study [69], Giergiel *et al* investigated a variant scenario of two mirrors wedging with the angle $\pi/4$, i.e., the two mirrors are located at $x = 0$ and $y = x$. In this way, they find it is possible to engineer two dimensional inseparable time crystalline structures of various geometries on the Möbius strip as shown in figure 6.3(b). The two mirrors wedge with a $\frac{\pi}{4}$ angle by setting one mirror in the vertical direction. The vertical mirror enforces additional condition $y \geqslant x$ for the atom. In the $\theta_x - \theta_y$ space, this restraint from the two mirrors becomes $(\theta_x - \theta_y)(\theta_x + \theta_y - 2\pi) \geqslant 0$, which makes the accessible region by half as marked green in the lower left panel. The solid white lines indicate the typical trajectory. The white dashed lines indicate the connection condition due to the collision with the vertical mirror. This defines a Möbius strip geometry as shown in the lower right panel of figure 6.3(b). Considering many interacting atoms, the authors show it is possible to realise a Lieb lattice model with a flat band and control long-range hopping of pairs of atoms in the model [69].

6.2.4.2 Space-time crystals

The systems discussed above to generate time crystals have no discrete symmetry in space. It is also possible to combine both spatial and temporal periodic structures simultaneously, which is called *space-time crystals* [7, 73]. In this way, the time crystalline space serves as a synthetic dimension, which can be harnessed to reveal the high-dimensional (e.g., six dimensional) quantum Hall effect.

In [74], the authors considered a particle in a 1D spatially periodic and periodically shaken potential, which is described by the following scaled single-particle Hamiltonian (in the rotating frame)

$$H_s(x, p_x, t) = p_x^2 + V_0 \sin^2 x + p_x \lambda \omega \sin \omega_d t, \tag{6.39}$$

where V_0 is the lattice potential depth and the last term results from the transition to the rotating frame. In the rest frame, the position of the optical lattice is harmonically modulated with a small amplitude λ and frequency ω_d. The driving frequency is chosen resonant with the motion of a particle, which means that the classical motion of a trapped particle with energy $E < V_0$ has oscillating frequency $\Omega(E) = \omega_d/s$ with integer $s > 1$. In the ith lattice site, the time-independent Hamiltonian in the vicinity of the resonant orbit characterised with the action I_s is given by

$$H_{\text{eff}}^{(i)} \approx \frac{P^2}{2m_{\text{eff}}} - V_{\text{eff}} \cos(s\Theta), \tag{6.40}$$

where $P = I - I_s$. Therefore, at each lattice site, a particle has a synthetic dimension with s sites due to the resonant orbit. In quantum dynamics, the eigenstates of Hamiltonian (6.40) are Bloch waves in the synthetic lattice and the corresponding quasienergies for $s \gg 1$ form energy bands. For the lowest energy band, we can define the Wannier states $W_{i,\alpha}(\Theta)$ which are localised in individual sites ($\alpha = 1, 2, \ldots, s$) of the effective potential $-V_{\text{eff}} \cos(s\Theta)$.

In the subspace spanned by the Wannier states, the total Floquet Hamiltonian describing the entire system beyond a single site takes the form of the tight-binding model

$$\mathcal{H} = -\frac{1}{2} \sum_{i,\alpha;j,\beta} J_{i,\alpha}^{j,\beta} a_{j,\beta}^\dagger a_{i,\beta}, \tag{6.41}$$

where $a_{i,\alpha}$ is the ladder operator acting on the lattice site (i, α). The tunnelling rate is

$$J_{i,\alpha}^{j,\beta} = -\frac{2\omega_d}{2\pi s} \int_0^{s2\pi/\omega_d} dt \langle W_{j,\beta} | (H_s - i\partial_t) | W_{i,\alpha} \rangle, \tag{6.42}$$

which describes the hopping transitions between sites of the optical lattice ($i \to j$) and between the synthetic lattice sites ($\alpha \to \beta$).

By introducing the photon-assisted tunnelling against a potential gradient along the temporal direction [74], one can realise a complex tunnelling amplitudes $J_{i,\alpha}^{i,\beta} = |J_{i,\alpha}^{i,\beta}| e^{i\phi_i}$, where the phase ϕ_i is a linear function of the site index i. When an atom tunnels around a plaquette of the 2D time-space crystal, it acquires a phase associated with a magnetic like flux. Therefore, the present approach reveal the 2D quantum Hall effect. The realisation of the topological 2D time-space crystal can be generalised to higher dimensions. By considering a particle in a 3D lattice vibrating in three orthogonal directions, one can realise a 6D time-space crystal and 6D quantum Hall effect quantified by the third Chern number [74–76].

6.3 Time crystals and phase space crystals

Time crystals and phase space crystals are closely related but different concepts. In figure 6.4, we schematically clarify the relationship between the two research fields of PSCs and FTCs. They both study the subharmonic modes that emerge in periodically driven systems. The subharmonic response at a fraction of driving frequency ω_d/n ($n > 1$), which can be determined by the time evolution of an observable, i.e.,

$$\begin{aligned}\langle O(t)\rangle_i &= A_i \cos(\omega_d t/n + \Theta_i)\\ &= X_i \cos(\omega_d t/n) + P_i \sin(\omega_d t/n),\end{aligned} \tag{6.43}$$

where i is the index of subharmonic modes. In the second line, we write the subharmonic modes in an equivalent form with quadratures (X_i, P_i). For a 1D system, the potential subharmonic modes after DTTS breaking are arranged periodically in their oscillating phases, i.e., $\Theta_i = 2\pi i/n$ ($i = 0, 2, \ldots, n-1$), see section 6.2.2. In the space of quadratures (phase space), these subharmonic modes are periodically distributed in a lattice structure along a circle as shown in figure 6.4. A concrete example is the power-law driven nonlinear oscillator shown in figure 3.1(a).

The subharmonic modes can also differ from both quadratures and form more complex 2D lattice structures in phase space, see the honeycomb crystal as shown in figure 6.4 and also the square, triangular and quasicrystal lattices discussed in section 3.1.2. We emphasise that one should distinguish the 2D phase space lattice from the effective 2D lattice potential of time crystals, cf. equation (6.30). Strictly speaking, the momentum direction is not an independent degree of freedom. The physical system to generate 2D phase space lattice is still a 1D system. In phase space lattices, the topological physics naturally emerges from the noncommutative

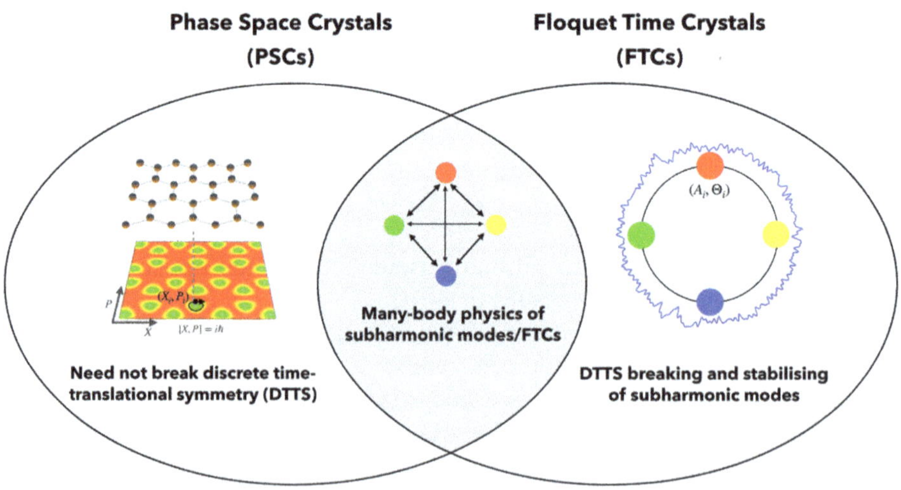

Figure 6.4. Relationship between PSCs and FTCs. FTCs focus on the DTTS breaking and the protection mechanisms (e.g., disorder) of subharmonic response modes. PSCs pay more attention on the periodically organised pattern in phase space and the interplay of potential (subharmonic) response modes which need not break DTTS.

geometry of phase space in quantum dynamics as discussed in section 3.3, or the symplectic geometry of phase space in classical dynamics as discussed in chapter 5.

Most studies of time crystals focus on the spontaneous breaking of DTTS and the protection mechanism of subharmonic modes in quantum many-body systems. As a contrast, the basic modes constructing a phase space crystal are not necessarily a many-body state. They can also be the single-particle modes that need not break DTTS. In fact, the kicking sequence to generate honeycomb phase space lattice has the same time period as the harmonic oscillator, see table 5.1. Phase space crystals focus on the many-body physics of these single-particle modes with consideration of their interactions. Especially, one should distinguish the phase space lattice, which is the effective single-particle Hamiltonian in phase space, and the phase space crystal, which is a many-body state of interacting single-particle modes. There is a trend to study the interplay of two or more time crystals, e.g., the AC Josephson effect between two superfluid time crystals [77]. The study of many-body physics of many interacting time crystals is coined as *condensed matter physics in time crystals* [31, 34, 78], a young research field still waiting to be explored.

References

[1] Wilczek F 2012 Quantum time crystals *Phys. Rev. Lett.* **109** 160401

[2] Shapere A and Wilczek F 2012 Classical time crystals *Phys. Rev. Lett.* **109** 160402

[3] Leff H S and Rex A F 2002 Maxwell's Demon 2: Entropy, Classical and Quantum Information *Computing* 1st edn (Boca Raton, FL: CRC Press)

[4] Feynman R P 1963 *The Feynman Lectures on Physics* Vol I (Reading, MA: Addison-Wesley) ch 46

[5] Bruno P 2013 Comment on 'quantum time crystals *Phys. Rev. Lett.* **110** 118901

[6] Kanamoto R, Saito H and Ueda M 2003 Stability of the quantized circulation of an attractive Bose-Einstein condensate in a rotating torus *Phys. Rev.* A **68** 043619

[7] Li T, Gong Z-X, Yin Z-Q, Yin H T Q, Zhang X, Duan P L-M and Zhang X 2012 Space-time crystals of trapped ions *Phys. Rev. Lett.* **109** 163001

[8] Nozières P 2013 Time crystals: can diamagnetic currents drive a charge density wave into rotation? *Europhys. Lett.* **103** 57008

[9] Bogoliubov N N 1970 *Lectures on Quantum Statistics* vol 2, 1st edn (New York: Gordon and Breach) p 1

[10] Bruno P 2013 Impossibility of spontaneously rotating time crystals: a no-go theorem *Phys. Rev. Lett.* **111** 070402

[11] Syrwid A, Zakrzewski J and Sacha K 2017 Time crystal behavior of excited eigenstates *Phys. Rev. Lett.* **119** 250602

[12] Autti S, Eltsov V B and Volovik G E 2018 Observation of a time quasicrystal and its transition to a superfluid time crystal *Phys. Rev. Lett.* **120** 215301

[13] Shapere A D and Wilczek F 2019 Regularizations of time-crystal dynamics *Proc. Natl Acad. Sci.* **116** 18772–6

[14] Wilczek F 2013 Superfluidity and space-time translation symmetry breaking *Phys. Rev. Lett.* **111** 250402

[15] Yoshii R, Takada S, Tsuchiya S, Marmorini G, Hayakawa H and Nitta M 2015 Fulde-Ferrell-Larkin-Ovchinnikov states in a superconducting ring with magnetic fields: phase diagram and the first-order phase transitions *Phys. Rev.* B **92** 224512

[16] Watanabe H and Oshikawa M 2015 Absence of quantum time crystals *Phys. Rev. Lett.* **114** 251603
[17] Lieb E H and Robinson D W 1972 The finite group velocity of quantum spin systems *Commun. Math. Phys.* **28** 251–7
[18] Khemani V, von Keyserlingk C W and Sondhi S L 2017 Defining time crystals via representation theory *Phys. Rev.* B **96** 115127
[19] Kozin V K and Kyriienko O 2019 Quantum time crystals from Hamiltonians with long-range interactions *Phys. Rev. Lett.* **123** 210602
[20] Sacha K 2015 Modeling spontaneous breaking of time-translation symmetry *Phys. Rev.* A **91** 033617
[21] Khemani V, Lazarides A, Moessner R and Sondhi S L 2016 Phase structure of driven quantum systems *Phys. Rev. Lett.* **116** 250401
[22] Else D V, Bauer B and Nayak C 2016 Floquet time crystals *Phys. Rev. Lett.* **117** 090402
[23] von Keyserlingk C W, Khemani V and Sondhi S L 2016 Absolute stability and spatiotemporal long-range order in Floquet systems *Phys. Rev.* B **94** 085112
[24] Yao N Y, Potter A C, Potirniche I-D and Vishwanath A 2017 Discrete time crystals: rigidity, criticality, and realizations *Phys. Rev. Lett.* **118** 030401
[25] Zhang J *et al* 2017 Observation of a discrete time crystal *Nature* **543** 217–20
[26] Choi S *et al* 2017 Observation of discrete time-crystalline order in a disordered dipolar many-body system *Nature* **543** 221–5
[27] Rovny J, Blum R L and Barrett S E 2018 Observation of discrete-time-crystal signatures in an ordered dipolar many-body system *Phys. Rev. Lett.* **120** 180603
[28] Rovny J, Blum R L and Barrett S E 2018 ^{31}P NMR study of discrete time-crystalline signatures in an ordered crystal of ammonium dihydrogen phosphate *Phys. Rev.* B **97** 184301
[29] Pal S, Nishad N, Mahesh T S and Sreejith G J 2018 Temporal order in periodically driven spins in star-shaped clusters *Phys. Rev. Lett.* **120** 180602
[30] Smits J, Liao L, Stoof H T C and van der Straten P 2018 Observation of a space-time crystal in a superfluid quantum gas *Phys. Rev. Lett.* **121** 185301
[31] Sacha K and Zakrzewski J 2017 Time crystals: a review *Rep. Prog. Phys.* **81** 016401
[32] Khemani V, Moessner R and Sondhi S L 2019 A brief history of time crystals arXiv:1910.10745
[33] Else D V, Monroe C, Nayak C and Yao N Y 2020 Discrete time crystals *Annu. Rev. Condens. Matter Phys.* **11** 467–99
[34] Guo L and Liang P 2020 Condensed matter physics in time crystals *New J. Phys.* **22** 075003
[35] Giergiel K, Kosior A, Hannaford P and Sacha K 2018 Time crystals: Analysis of experimental conditions *Phys. Rev.* A **98** 013613
[36] Sreejith G J, Lazarides A and Moessner R 2016 Parafermion chain with $2\pi/k$ Floquet edge modes *Phys. Rev.* B **94** 045127
[37] Surace F M, Russomanno A, Dalmonte M, Silva A, Fazio R and Iemini F 2019 Floquet time crystals in clock models *Phys. Rev.* B **99** 104303
[38] Pizzi A, Knolle J and Nunnenkamp A 2020 Period-n discrete time crystals and quasicrystals with ultracold bosons *Phys. Rev. Lett.* **123** 150601
[39] Pizzi A, Knolle J and Nunnenkamp A 2021 Higher-order and fractional discrete time crystals in clean long-range interacting systems *Nat. Commun.* **12** 2341
[40] Matus P and Sacha K 2019 Fractional time crystals *Phys. Rev.* A **99** 033626
[41] Faraday M 1831 On a peculiar class of acoustical figures; and on certain forms assumed by groups of particles upon vibrating elastic surfaces *Philos. Trans. R. Soc. Lond.* **121** 299–340

[42] Lazarides A, Das A and Moessner R 2014 Equilibrium states of generic quantum systems subject to periodic driving *Phys. Rev.* E **90** 012110

[43] D'Alessio L and Rigol M 2014 Long-time behavior of isolated periodically driven interacting lattice systems *Phys. Rev.* X **4** 041048

[44] Ponte P, Chandran A, Papić Z and Abanin D A 2015 Periodically driven ergodic and many-body localized quantum systems *Ann. Phys., NY* **353** 196–204

[45] Nandkishore R and Huse D A 2015 Many-body localization and thermalization in quantum statistical mechanics *Annu. Rev. Condens. Matter Phys.* **6** 15–38

[46] Ponte P, Papić Z, Huveneers F and Abanin D A 2015 Many-body localization in periodically driven systems *Phys. Rev. Lett.* **114** 140401

[47] Lazarides A, Das A and Moessner R 2015 Fate of many-body localization under periodic driving *Phys. Rev. Lett.* **115** 030402

[48] Abanin D A, De Roeck W and Huveneers F 2016 Theory of many-body localization in periodically driven systems *Ann. Phys., NY* **372** 1–11

[49] Pal A and Huse D A 2010 Many-body localization phase transition *Phys. Rev.* B **82** 174411

[50] Abanin D A, De Roeck W and Huveneers F 2015 Exponentially slow heating in periodically driven many-body systems *Phys. Rev. Lett.* **115** 256803

[51] Abanin D A, De Roeck W, Ho W W and Huveneers F 2017 Effective Hamiltonians, prethermalization, and slow energy absorption in periodically driven many-body systems *Phys. Rev.* B **95** 014112

[52] Mori T, Kuwahara T and Saito K 2016 Rigorous bound on energy absorption and generic relaxation in periodically driven quantum systems *Phys. Rev. Lett.* **116** 120401

[53] Kuwahara T, Mori T and Saito K 2016 Floquet-Magnus theory and generic transient dynamics in periodically driven many-body quantum systems *Ann. Phys., NY* **367** 96–124

[54] Russomanno A, Iemini F, Dalmonte M and Fazio R 2017 Floquet time crystal in the Lipkin-Meshkov-Glick model *Phys. Rev.* B **95** 214307

[55] Huang B, Wu Y-H and Liu W V 2018 Clean Floquet time crystals: models and realizations in cold atoms *Phys. Rev. Lett.* **120** 110603

[56] Ho W W, Choi S, Lukin M D and Abanin D A 2017 Critical time crystals in dipolar systems *Phys. Rev. Lett.* **119** 010602

[57] Barfknecht R E, Rasmussen S E, Foerster A and Zinner N T 2019 Realizing time crystals in discrete quantum few-body systems *Phys. Rev.* B **99** 144304

[58] Else D V, Bauer B and Nayak C 2017 Prethermal phases of matter protected by time-translation symmetry *Phys. Rev.* X **7** 011026

[59] Luitz D J, Moessner R, Sondhi S L and Khemani V 2020 Prethermalization without temperature *Phys. Rev.* X **10** 021046

[60] Kuramoto Y and Nishikawa I 1987 Statistical macrodynamics of large dynamical systems. case of a phase transition in oscillator communities *J. Stat. Phys.* **49** 569–605

[61] Lindsay J, Field R J and Burger M 1985 *Oscillations and Traveling Waves in Chemical Systems* (New York: Wiley)

[62] Heo M-S, Kim Y, Kim K, Moon G, Lee J, Noh H-R, Dykman M I and Jhe W 2010 Ideal mean-field transition in a modulated cold atom system *Phys. Rev.* E **82** 031134

[63] Kim K, Heo M-S, Lee K-H, Jang K, Noh H-R, Kim D and Jhe W 2006 Spontaneous symmetry breaking of population in a nonadiabatically driven atomic trap: an Ising-class phase transition *Phys. Rev. Lett.* **96** 150601

[64] Svensson I-M, Bengtsson A, Krantz P, Bylander J, Shumeiko V and Delsing P 2017 Period-tripling subharmonic oscillations in a driven superconducting resonator *Phys. Rev.* B **96** 174503

[65] Svensson I-M, Bengtsson A, Bylander J, Shumeiko V and Delsing P 2018 Period multiplication in a parametrically driven superconducting resonator *Appl. Phys. Lett.* **113** 022602

[66] Wustmann W and Shumeiko V 2019 Parametric effects in circuit quantum electrodynamics *Low Temp. Phys.* **45** 848–69

[67] Chinzei K and Ikeda T N 2020 Time crystals protected by Floquet dynamical symmetry in Hubbard models *Phys. Rev. Lett.* **125** 060601

[68] Delande D, Morales-Molina L and Sacha K 2017 Three-dimensional localized-delocalized Anderson transition in the time domain *Phys. Rev. Lett.* **119** 230404

[69] Giergiel K, Kuroś A, Kosior A and Sacha K 2021 Inseparable time-crystal geometries on the Möbius strip *Phys. Rev. Lett.* **127** 263003

[70] Giergiel K, Miroszewski A and Sacha K 2018 Time crystal platform: from quasicrystal structures in time to systems with exotic interactions *Phys. Rev. Lett.* **120** 140401

[71] Giergiel K, Kuroś A and Sacha K 2019 Discrete time quasicrystals *Phys. Rev.* B **99** 220303

[72] Buchleitner A, Delande D and Zakrzewski J 2002 Nondispersive wave packets in periodically driven quantum systems *Phys. Rep.* **368** 409–547

[73] Träger N *et al* 2021 Real-space observation of magnon interaction with driven space-time crystals *phys. Rev. Lett.* **126** 057201

[74] Žlabys G, Fan C-h, Anisimovas E and Sacha K 2021 Six-dimensional time-space crystalline structures *Phys. Rev.* B **103** L100301

[75] Petrides I, Price H M and Zilberberg O 2018 Six-dimensional quantum hall effect and three-dimensional topological pumps *Phys. Rev.* B **98** 125431

[76] Lee C H, Wang Y, Chen Y and Zhang X 2018 Electromagnetic response of quantum hall systems in dimensions five and six and beyond *Phys. Rev.* B **98** 094434

[77] Autti S, Heikkinen P J, Mäkinen J T, Volovik G E, Zavjalov V V and Eltsov V B 2021 AC Josephson effect between two superfluid time crystals *Nat. Mater.* **20** 171–4

[78] Sacha K 2020 *Condensed Matter Physics in the Time Dimension* (Cham Springer) pp 173–235

IOP Publishing

Phase Space Crystals
Condensed matter in dynamical systems
Lingzhen Guo

Appendix A

Phase space lattices

A.1 General theory

We discuss how to synthesise arbitrary lattice structures in phase space via multiple stroboscopic lattices. We start from the following generalised model of a kicked harmonic oscillator

$$H_s = \frac{1}{2}(x^2 + p^2) + \sum_{n\in\mathbb{Z}} \sum_q K_q \cos(k_q x - \phi_q)\delta\left(\frac{t}{\tau} - \theta_q - n\right), \tag{A.1}$$

where q represents the kicking sequence of stroboscopic lattices with tunable intensity K_q, wave vector k_q and phase ϕ_q at different time instances $t = \tau(n + \theta_q)$ with $n \in \mathbb{Z}$. To simplify the discussion, we first consider a single kick sequence, i.e.,

$$H_s = \frac{1}{2}(x^2 + p^2) + K_q \cos(k_q x - \phi_q)\delta\left(\frac{t}{\tau} - \theta_q - n\right). \tag{A.2}$$

We transform the above Hamiltonian into a rotating frame with kicking frequency $2\pi/\tau$ using the generating function of the second kind

$$G_2(x, P, t) = \frac{xP}{\cos(2\pi t/\tau)} - \frac{1}{2}x^2 \tan\left(\frac{2\pi}{\tau}t\right) - \frac{1}{2}P^2 \tan\left(\frac{2\pi}{\tau}t\right). \tag{A.3}$$

The corresponding canonical transformation is given by, cf. equation (2.47),

$$p = \frac{\partial G_2}{\partial x}, \quad X = \frac{\partial G_2}{\partial P}, \quad H_s(X, P, t) = H_s(x, p, t) + \frac{\partial G_2}{\partial t},$$

which results in the transformation of phase space coordinates

$$x = P\sin\left(\frac{2\pi t}{\tau}\right) + X\cos\left(\frac{2\pi t}{\tau}\right), \quad p = P\cos\left(\frac{2\pi t}{\tau}\right) - X\sin\left(\frac{2\pi t}{\tau}\right) \tag{A.4}$$

and the transformed Hamiltonian

$$\begin{aligned} H_s(X, P, t) &= \frac{1}{2}\delta\omega(X^2 + P^2) \\ &+ K_q \cos\left[k_q\left(P\sin\frac{2\pi t}{\tau} + X\cos\frac{2\pi t}{\tau}\right) - \phi_q\right]\delta\left(\frac{t}{\tau} - \theta_q - n\right). \end{aligned} \tag{A.5}$$

Here, $\delta\omega \equiv 1 - 2\pi/\tau$ is the detuning between kicking and harmonic oscillation.

For weak resonant driving ($|K_q| \ll 1, \tau = 2\pi$), the single-particle dynamics can be separated by the fast harmonic oscillation and the low motion of its quadratures (X, P). The effective slow dynamics of quadratures (X, P) is given by the lowest-order Magnus expansion (see equation (2.134)), i.e., the time average of $H_s(X, P, t)$ in one kicking time period,

$$\begin{aligned} H_s(X, P) &= \frac{1}{\tau}\int_0^\tau H_s(X, P, t) \\ dt &= K_q \cos\left[k_q\left(P\sin 2\pi\theta_q + X\cos 2\pi\theta_q\right) - \phi_q\right]. \end{aligned} \tag{A.6}$$

The above derivation is valid for classical dynamics. For quantum dynamics, the result has the same form with replacing X and P by their operators. Including all the kicks in equation (A.1), we obtain the general form of the phase space lattice Hamiltonian

$$H_s(X, P) = \sum_q K_q \cos\left[k_q\left(P\sin 2\pi\theta_q + X\cos 2\pi\theta_q\right) - \phi_q\right]. \tag{A.7}$$

In principle, any arbitrary lattice Hamiltonian in phase space can be synthesised by multiple stroboscopic lattices.

A.2 Honeycomb phase space lattice

For the honeycomb lattice considered in our work, we can get the desired driving parameters by decomposing the honeycomb lattice into a series of cosine functions, i.e.,

$$
\begin{aligned}
H_s(X, P) = {}& 16\Lambda \sin^2\left(\frac{1}{2}X + \frac{1}{2\sqrt{3}}P\right)\sin^2\left(-\frac{1}{2}X + \frac{1}{2\sqrt{3}}P\right)\sin^2\left(\frac{1}{\sqrt{3}}P\right) \\
= {}& \Lambda\left(\frac{3}{2} + \cos[2(X\cos 0 + P\sin 0)]\right. \\
& + \cos\left[2\left(X\cos\frac{\pi}{3} + P\sin\frac{\pi}{3}\right)\right] \\
& + \cos\left[2\left(X\cos\frac{2\pi}{3} + P\sin\frac{2\pi}{3}\right)\right] \\
& + \frac{1}{2}\cos\left[\frac{4}{\sqrt{3}}\left(X\cos\frac{\pi}{6} + P\sin\frac{\pi}{6}\right) - \pi\right] \\
& + \frac{1}{2}\cos\left[\frac{4}{\sqrt{3}}\left(X\cos\frac{5\pi}{6} + P\sin\frac{5\pi}{6}\right) - \pi\right] \\
& + \frac{1}{2}\cos\left[\frac{4}{\sqrt{3}}\left(X\cos\frac{3\pi}{2} + P\sin\frac{3\pi}{2}\right) - \pi\right] \\
& + \cos\left[\frac{2}{\sqrt{3}}\left(X\cos\frac{\pi}{2} + P\sin\frac{\pi}{2}\right) - \pi\right] \\
& + \cos\left[\frac{2}{\sqrt{3}}\left(X\cos\frac{7\pi}{6} + P\sin\frac{7\pi}{6}\right) - \pi\right] \\
& \left. + \cos\left[\frac{2}{\sqrt{3}}\left(X\cos\frac{11\pi}{6} + P\sin\frac{11\pi}{6}\right) - \pi\right]\right).
\end{aligned}
\tag{A.8}
$$

Comparing the above expansion to equation (A.7), we can generate the honeycomb phase space Hamiltonian (up to a constant $3\Lambda/2$) by choosing the kicking parameters $K_q = \Lambda$, $k_q = 2$ and $\phi_q = 0$ for $\theta_q = 0, 2/12, 4/12$; $K_q = \Lambda/2$, $k_q = 4/\sqrt{3}$ and $\phi_q = \pi$ for $\theta_q = 1/12, 5/12, 9/12$; $K_q = \Lambda$, $k_q = 2/\sqrt{3}$ and $\phi_q = \pi$ for $\theta_q = 3/12, 7/12, 11/12$.

The extrema of $H_s(X, P)$ represent stable points of the classical dynamics. The small vibrations about these stable points have frequency $\omega_0 = -9\Lambda$ from the linear expansion of equation (A.8). We rewrite the honeycomb lattice Hamiltonian in the form of

$$
H_s(\boldsymbol{Z}) = -\frac{16}{9}\omega_0 \prod_{n=1}^{3} \sin^2\left(\frac{1}{2}v_n \cdot \boldsymbol{Z}\right), \tag{A.9}
$$

where we have defined the vector $\boldsymbol{Z} \equiv (X, P)$ and three ancillary vectors $v_1 = (1, \frac{1}{\sqrt{3}})$, $v_2 = (-1, \frac{1}{\sqrt{3}})$, $v_3 = (0, -\frac{2}{\sqrt{3}})$. The two sublattices of the phase space honeycomb lattice Hamiltonian (A.9) have the same vibration frequency at their lattice sites. One can also tune the on-site frequencies of two sublattices with the following phase space lattice Hamiltonian

$$H_s^{\Delta} = -\frac{2\Delta}{\sqrt{3}} \sum_{n=1}^{3} \sin(\boldsymbol{v}_n \cdot \boldsymbol{Z}) = \frac{2\Delta}{\sqrt{3}} \sum_{n=1}^{3} \cos\left(\boldsymbol{v}_n \cdot \boldsymbol{Z} - \frac{3\pi}{2}\right), \tag{A.10}$$

which can be implemented by additional stroboscopic lattices with kicking parameters $K_q = 2\Delta/\sqrt{3}$, $k_q = 2/\sqrt{3}$ and $\phi_q = 3\pi/2$ for $\theta_q = 1/12, 5/12, 9/12$. The on-site frequencies of two sublattices are then $\omega_0 \pm 2\Delta$ with the parameter Δ which we call the *on-site detuning*. The total single-particle Hamiltonian of the honeycomb lattice including the on-site detuning terms is given by

$$H_s(\boldsymbol{Z}) = -\frac{16}{9}\omega_0 \prod_{n=1}^{3} \sin^2\left(\frac{1}{2}\boldsymbol{v}_n \cdot \boldsymbol{Z}\right) - \frac{2\Delta}{\sqrt{3}} \sum_{n=1}^{3} \sin(\boldsymbol{v}_n \cdot \boldsymbol{Z}). \tag{A.11}$$

IOP Publishing

Phase Space Crystals
Condensed matter in dynamical systems
Lingzhen Guo

Appendix B

Phase space interactions

B.1 Direct and exchange interactions

We provide the analytical expressions for the classical direct interaction and quantum exchange interaction introduced in section 4.2.2.

First, we calculate the matrix element of $U(\hat{R})$ for two given functions $f(X_1, X_2)$ and $g(X_1, X_2)$ as follows

$$
\begin{aligned}
&\langle f(X_1, X_2)|U(\hat{R})|g(X_1, X_2)\rangle \\
&= \langle f(X_c + \frac{1}{2}\Delta X, X_c - \frac{1}{2}\Delta X)|U(\hat{R})|g(X_c + \frac{1}{2}\Delta X, X_c - \frac{1}{2}\Delta X)\rangle \\
&= \sum_N U(R_N) \int dC \langle f(X_c + \frac{1}{2}\Delta X, X_c - \frac{1}{2}\Delta X)|N, C\rangle \\
&\times \langle N, C|g(X_c + \frac{1}{2}\Delta X, X_c - \frac{1}{2}\Delta X)\rangle \\
&= \sum_N U(R_N) \int dC \langle f(C + \frac{1}{2}\Delta X, C - \frac{1}{2}\Delta X)|\Phi_N(\Delta X)\rangle \\
&\times \langle \Phi_N(\Delta X)|g(C + \frac{1}{2}\Delta X, C - \frac{1}{2}\Delta X)\rangle.
\end{aligned}
\tag{B.1}
$$

Here, we have used the property of $\hat{X}_c|C\rangle = C|C\rangle$ and the result

$$\langle f(X_c)|C\rangle = \int f(X_c)\delta(X_c - C)dX_c = f(C)$$

for any $f(X_c)$ in the representation of coordinate of centre of mass.

Then, we apply the result equation (B.1) to calculate U and J defined by equation (4.45), i.e., the direct integral

$$
\begin{aligned}
U &= \langle \varphi(X_1)\phi(X_2)|U(\hat{R})|\varphi(X_1)\phi(X_2)\rangle \\
&= \sum_N U(R_N) \int dC \langle \varphi(C + \frac{1}{2}\Delta X)\phi(C - \frac{1}{2}\Delta X)|\Phi_N(\Delta X)\rangle \\
&\quad \times \langle \Phi_N(\Delta X)|\varphi(C + \frac{1}{2}\Delta X)\phi(C - \frac{1}{2}\Delta X)\rangle \\
&= \sum_N U(R_N) \int dC\, |\langle \varphi(C + \frac{1}{2}\Delta X)\phi(C - \frac{1}{2}\Delta X)|\Phi_N(\Delta X)\rangle|^2 \\
&\equiv \sum_N U(R_N) I_N,
\end{aligned} \tag{B.2}
$$

and the exchange integral

$$
\begin{aligned}
J &= \langle \varphi(X_1)\phi(X_2)|U(\hat{R})|\phi(X_1)\varphi(X_2)\rangle \\
&= \sum_N U(R_N) \int dC \langle \varphi(C + \frac{1}{2}\Delta X)\phi(C - \frac{1}{2}\Delta X)|\Phi_N(\Delta X)\rangle \\
&\quad \times \langle \Phi_N(\Delta X)|\phi(C + \frac{1}{2}\Delta X)\varphi(C - \frac{1}{2}\Delta X)\rangle \\
&= \sum_N (-1)^N U(R_N) \\
&\quad \times \int dC\, |\langle \varphi(C + \frac{1}{2}\Delta X)\phi(C - \frac{1}{2}\Delta X)|\Phi_N(\Delta X)\rangle|^2 \\
&\equiv \sum_N (-1)^N U(R_N) I_N.
\end{aligned} \tag{B.3}
$$

Here, the overlap integral I_N is given by

$$
I_N \equiv \int dC\, |\langle \varphi(C + \frac{1}{2}\Delta X)\phi(C - \frac{1}{2}\Delta X)|\Phi_N(\Delta X)\rangle|^2. \tag{B.4}
$$

B.2 Overlap integral of coherent states

Now, we assume the two states $\varphi(X)$ and $\phi(X)$ are two displaced squeezed coherent states described by

$$
\begin{cases}
\varphi(X) = \psi_0(X - r_m) = \left(\dfrac{\beta}{\sqrt{\pi}}\right)^{\frac{1}{2}} e^{-\frac{1}{2}\beta^2(X - r_m)^2} \\
\phi(X) = \psi_0(X + r_m) = \left(\dfrac{\beta}{\sqrt{\pi}}\right)^{\frac{1}{2}} e^{-\frac{1}{2}\beta^2(X + r_m)^2}.
\end{cases} \tag{B.5}
$$

The product state $\varphi(X_1)\phi(X_2) = \varphi(C + \frac{1}{2}\Delta X)\phi(C - \frac{1}{2}\Delta X)$ can be calculated from equation (B.5)

$$\varphi(X_1)\phi(X_2) = \frac{\beta}{\sqrt{\pi}} e^{-\beta^2 C^2} e^{-\beta^2\left(\frac{1}{2}\Delta X - r_m\right)^2}. \tag{B.6}$$

From equation (B.4), we obtain the following overlap integral

$$\begin{aligned} I_N &= \frac{\beta^2}{\pi}\int_{-\infty}^{+\infty} e^{-2\beta^2 C^2} dC\, |\langle e^{-\beta^2\left(\frac{1}{2}\Delta X - r_m\right)^2} |\Phi_N(\Delta X)\rangle|^2 \\ &= \left| \langle \Phi_N(\Delta X) | \left(\frac{\beta}{\sqrt{2\pi}}\right)^{\frac{1}{2}} e^{-\frac{1}{2}\left(\frac{\beta}{\sqrt{2}}\right)^2 \left(\Delta X - 2r_m\right)^2} \rangle \right|^2. \end{aligned} \tag{B.7}$$

The displacement operator $\hat{D}_\gamma = \exp(\gamma \hat{a}^\dagger - \gamma^* \hat{a})$ has the property $\hat{D}_\gamma^\dagger \hat{a} \hat{D}_\gamma = \hat{a} + \gamma$. We further introduce the squeezing operator

$$\hat{S}_\xi \equiv \exp\left[\frac{1}{2}(\xi^* \hat{a}^2 - \xi \hat{a}^{\dagger 2})\right]$$

with parameter $\xi = re^{i\theta}$. The squeezing operator has the following property

$$\hat{S}_\xi^\dagger \hat{a} \hat{S}_\xi = v\hat{a} + u\hat{a}^\dagger$$

with the squeezing parameters $v = \cosh r$, $u = -e^{i\theta}\sinh r$. Inversely, the parameter $\xi = re^{i\theta}$ is related to v and u via

$$\begin{cases} r &= \operatorname{arccosh}(v) = \ln(v + \sqrt{v^2 - 1}) \\ e^{i\theta} &= -u/\sinh r. \end{cases} \tag{B.8}$$

Using operators $\hat{D}_\gamma$ and $\hat{S}_\xi$, we write the displaced squeezed state in equation (B.7) as

$$\left| \left(\frac{\beta}{\sqrt{2\pi}}\right)^{\frac{1}{2}} e^{-\frac{1}{2}\left(\frac{\beta}{\sqrt{2}}\right)^2 (\Delta X - 2r_m)^2} \right\rangle = \hat{D}_{-\gamma}\hat{S}_{-\xi}|0\rangle \equiv |-\gamma, -\xi\rangle, \tag{B.9}$$

where the displacement parameter is $\gamma = -r_m/\sqrt{\lambda}$ and the squeezing parameters are

$$v = \frac{1}{2}\left(\sqrt{\lambda}\,\beta + \frac{1}{\sqrt{\lambda}\,\beta}\right), \quad u = \frac{1}{2}\left(\sqrt{\lambda}\,\beta - \frac{1}{\sqrt{\lambda}\,\beta}\right). \tag{B.10}$$

Using the formula (7.81) in the book of *Introductory Quantum Optics* by Christopher Gerry and Peter Knight, the overlap integral $I_N = |\langle N| -\gamma, -\xi\rangle|^2$ is given by

$$\begin{aligned} I_N = \frac{\left(\frac{1}{2}\tanh r\right)^N}{N!\cosh r} \left| H_N\left[\frac{\gamma^* e^{i\theta}\sinh r - \gamma\cosh r}{\sqrt{e^{i(\theta+\pi)}\sinh(2r)}}\right]\right|^2 \\ \exp\left[-|\gamma|^2 + \frac{1}{2}(\gamma^{*2}e^{i\theta} + \gamma^2 e^{-i\theta})\tanh r\right]. \end{aligned} \tag{B.11}$$

Given the parameters r_m and β, we can calculate U_c and U_e from equations (4.46), (B.10) and (B.11).

The standard coherent state, whose squeezing parameters are v = 1 and $u = 0$, can be obtained by choosing $\beta = 1/\sqrt{\lambda}$ in equation (B.10). From equation (B.11), the overlap integral of two standard coherent states I_N^s can be calculated

$$I_N^s = \frac{|\gamma|^{2N}}{N!} e^{-|\gamma|^2} = \frac{1}{N!}\left(\frac{R^2}{4\lambda}\right)^N \exp\left(-\frac{R^2}{4\lambda}\right). \tag{B.12}$$

Here, $R \equiv 2r_m = 2\gamma\sqrt{\lambda}$ is the distance between the centres of two coherent states in phase space. The quantity R is different from the quantised phase space distance R_N. For two overlapped coherent states, their distance R is zero but R_N is always positive as shown by equation (4.39).

B.3 Hard-sphere interaction

We derive $U(R_N)$ for hard-sphere interaction in equation (4.62). Assuming the interaction potential between two atoms is $V(x_1 - x_2)$, the eigenequation of energy is given by

$$\left[-\frac{\lambda^2}{2}\frac{\partial^2}{\partial x_1^2} - \frac{\lambda^2}{2}\frac{\partial^2}{\partial x_2^2} + \frac{1}{2}\left(x_1^2 + x_2^2\right) + V(x_1 - x_2)\right]\Psi(x_1, x_2) = E_T\Psi(x_1, x_2). \tag{B.13}$$

Herr, E_T is the total energy. We introduce the coordinate of central of mass $x_c \equiv (x_1 + x_2)/2$ and the relative coordinate $x \equiv x_1 - x_2$. By separating the eigenstate into a product state $\Psi = \phi(x_c)\psi(x)$, we have the eigenequation for the motion of centre of mass

$$\left(-\frac{\lambda^2}{2M}\frac{\partial^2}{\partial x_c^2} + \frac{1}{2}Mx_c^2\right)\phi(x_c) = E_c\phi(x_c) \tag{B.14}$$

with $M = 2$ the total mass and E_c the energy of centre of mass motion. The eigenequation for the relative motion is

$$\left[-\frac{\lambda^2}{2\mu}\frac{\partial^2}{\partial x^2} + \frac{1}{2}\mu x^2 + V(x)\right]\psi(x) = E\psi(x) \tag{B.15}$$

with $\mu = 1/2$ the reduced mass, $E = E_T - E_c$ the energy of relative motion. The solutions of equation (B.14) are just the harmonic motions. We now try to find the solutions of equation (B.15). Without interaction $V(x_1 - x_2)$, the eigenproblem is determined by $H_r\phi_n(x) = E_n\phi_n(x)$ with

$$H_r \equiv -\frac{\lambda^2}{2\mu}\frac{\partial^2}{\partial x^2} + \frac{1}{2}\mu x^2.$$

The eigenstates are given by

$$\psi_n(x) = \left(\frac{\zeta}{\sqrt{\pi}2^n n!}\right)^{\frac{1}{2}} H_n(\zeta x)e^{-\frac{1}{2}\zeta^2 x^2}, \tag{B.16}$$

with the parameter $\zeta = \sqrt{1/(2\lambda)}$. For the hard-sphere interaction, i.e., $V(x_1 - x_2) = +\infty$ for $|x_1 - x_2| < 2\sigma$ and $V(x_1 - x_2) = 0$ for $|x_1 - x_2| > 2\sigma$, the boundary condition requires that wavefunction must be zero at $x \in [-\sigma, \sigma]$. For odd integer n, we assume the approximate eigenstates are just repulsed outside the hard-sphere region, i.e.,

$$\phi_{2n+1}(x) = \begin{cases} \psi_{2n+1}(x - 2\sigma) & x \geqslant 2\sigma \\ 0 & -2\sigma < x < 2\sigma \quad n \in \mathbb{N}. \\ \psi_{2n+1}(x + 2\sigma) & x < -2\sigma \end{cases} \tag{B.17}$$

For even integer n, the wave functions $\phi_n(x)$, however, does not satisfy the hard-core boundary condition and the continuity condition. Therefore, we construct the symmetric eigenstates from antisymmetric states

$$\phi_{2n}(x) = \begin{cases} \psi_{2n+1}(x - 2\sigma) & x \geqslant 2\sigma \\ 0 & -2\sigma < x < 2\sigma \quad n \in \mathbb{N}. \\ -\psi_{2n+1}(x + 2\sigma) & x < -2\sigma \end{cases} \tag{B.18}$$

The energy levels to the first order correction are

$$\begin{aligned} &\langle \phi_{2n} | H_r | \phi_{2n} \rangle = \langle \phi_{2n+1} | H_r | \phi_{2n+1} \rangle \\ &= \frac{1}{2} \int_{-\infty}^{-2\sigma} dx\, \psi_{2n+1}^*(x + 2\sigma) \left(-\frac{\lambda^2}{\mu} \frac{\partial^2}{\partial x^2} \cdot + \mu x^2 \right) \psi_{2n+1}(x + 2\sigma) \\ &+ \frac{1}{2} \int_{2\sigma}^{\infty} dx\, \psi_{2n+1}^*(x - 2\sigma) \left(-\frac{\lambda^2}{\mu} \frac{\partial^2}{\partial x^2} + \mu x^2 \right) \psi_{2n+1}(x - 2\sigma) \\ &= \int_{-\infty}^{\infty} dx\, \psi_{2n+1}^*(x) H_r \psi_{2n+1}(x) + 2\mu^2\sigma^2 \int_{-\infty}^{\infty} dx\, \psi_{2n+1}^*(x) \psi_{2n+1}(x) \\ &+ 4\mu a \int_0^{\infty} dx\, \psi_{2n+1}^*(x)\, x\, \psi_{2n+1}(x) \\ &= \lambda(2n + 1) + \frac{1}{2}\sigma^2 + \frac{\sigma}{\zeta} \frac{1}{(2n + 1)! 2^{2n+1} \sqrt{\pi}} \\ &\times \sum_{k,l=0}^{\left[n+\frac{1}{2}\right],\left[n+\frac{1}{2}\right]} \frac{(-1)^{k+l} [(2n + 1)!]^2 (2n + 1 - k - l)!}{k! l! (2n + 1 - 2k)! (2n + 1 - 2l)!} 2^{2(2n+1-k-l)}, \end{aligned} \tag{B.19}$$

where we have used $\mu = 1/2$ in the last step. In fact, one can prove that $U(R_N)$ is the first order correction, from the weak interaction, to the Nth energy level of the harmonic trapping potential. Relabelling $N \equiv 2n + 1$, we have from equation (B.19)

$$U(R_N) = \frac{\sigma\sqrt{2\lambda/\pi}}{2^{-N} N!} \sum_{k,l=0}^{\left[\frac{N}{2}\right]\left[\frac{N}{2}\right]} \frac{(-1)^{k+l} (N!)^2 (N - k - l)!}{2^{2k+2l} k! l! (N - 2k)! (N - 2l)!}, \tag{B.20}$$

which is exactly equation (4.63) in section 4.2.3.

IOP Publishing

Phase Space Crystals

Condensed matter in dynamical systems

Lingzhen Guo

Appendix C

Phase space crystal vibrations

C.1 Hamiltonian of phase space crystal vibrations

We expand the total Hamiltonian H around the equilibrium positions of atoms to the second order (up to a constant)

$$H = \frac{1}{2}\sum_{\alpha,\beta}\sum_{L,L'}\eta_{\alpha\beta}^{LL'}u_L^\alpha u_{L'}^\beta \tag{C.1}$$

with the matrix $\eta_{\alpha\beta}^{LL'}$ given by

$$\eta_{\alpha\beta}^{LL'} = \left.\frac{\partial^2 H}{\partial u_L^\alpha \partial u_{L'}^\beta}\right|_0 = T_{\alpha\beta}^{LL'} + \Phi_{\alpha\beta}^{LL'}. \tag{C.2}$$

We call $\eta_{\alpha\beta}^{LL'}$ the *phase space force matrix*, which means the resultant force along the α direction exerted on the Lth atom due to the unit displacement along β direction of the L' th atom. The contribution to $\eta_{\alpha\beta}^{LL'}$ from the single-particle Hamiltonian (A.11) is

$$T_{\alpha\beta}^{LL'} = \left.\frac{\partial^2 H_s}{\partial u_L^\alpha \partial u_{L'}^\beta}\right|_0 = \left.\frac{\partial^2 H_s}{\partial u_L^\alpha \partial u_L^\beta}\right|_0 \delta_{LL'} = T_{\alpha\beta}^{LL}\delta_{LL'} = (\omega_0 \pm 2\Delta)\delta_{\alpha\beta}\delta_{LL'}. \tag{C.3}$$

The contribution from the interaction part can be obtained by calculating the derivative of interaction potential

$$\frac{\partial \Phi}{\partial Z_L^\alpha} = \sum_{L' \neq L}\frac{U'(R_{LL'})}{R_{LL'}}[(X_L - X_{L'})\delta_{X\alpha} + (P_L - P_{L'})\delta_{P\alpha}]. \tag{C.4}$$

For two different atoms at different lattice sites ($L \neq L'$), we have the second derivative

$$\begin{aligned}\Phi_{\alpha\beta}^{LL'} &= \frac{\partial^2 \Phi}{\partial Z_L^\alpha \partial Z_{L'}^\beta} \\ &= \frac{U'(R_{LL'})}{R_{LL'}}\left(-\delta_{X\alpha}\delta_{X\beta} - \delta_{P\alpha}\delta_{P\beta}\right) + \frac{U''(R_{LL'})R_{LL'} - U'(R_{LL'})}{R_{LL'}^3} \\ &\times\left[-(X_L - X_{L'})\delta_{X\beta} - (P_L - P_{L'})\delta_{P\beta}\right][(X_L - X_{L'})\delta_{X\alpha} + (P_L - P_{L'})\delta_{P\alpha}].\end{aligned} \tag{C.5}$$

For the atom at the single lattice site ($L = L'$), we have the second derivative

$$\begin{aligned}\Phi_{\alpha\beta}^{LL} &= \frac{\partial^2 \Phi}{\partial Z_L^\alpha \partial Z_L^\beta} \\ &= \sum_{L' \neq L} \frac{U'(R_{LL'})}{R_{LL'}}\left(\delta_{X\alpha}\delta_{X\beta} + \delta_{P\alpha}\delta_{P\beta}\right) + \frac{U''(R_{LL'})R_{LL'} - U'(R_{LL'})}{R_{LL'}^3} \\ &\times\left[(X_L - X_{L'})\delta_{X\beta} + (P_L - P_{L'})\delta_{P\beta}\right][(X_L - X_{L'})\delta_{X\alpha} + (P_L - P_{L'})\delta_{P\alpha}] \\ &= \sum_{L' \neq L} (-1) \times \frac{\partial^2 \Phi}{\partial Z_{L'}^\beta \partial Z_L^\alpha}.\end{aligned} \tag{C.6}$$

We introduce the complex field at each lattice site $a_L(t) \equiv \frac{1}{\sqrt{2\lambda}}[u_L^X(t) + iu_L^P(t)]$, where $\lambda = \hbar k^2/m\omega_{ax}$ is the dimensionless Planck constant, and obtain the Hamiltonian from equation (C.1)

$$\mathcal{H}/\lambda = \sum_{L,L'} h_{LL'}a_L^\dagger a_{L'} + \frac{1}{2}g_{LL'}a_L^\dagger a_{L'}^\dagger + \frac{1}{2}g_{L'L}^* a_L a_{L'}. \tag{C.7}$$

Here, h_{LL} represents the on-site energy of one excitation on the Lth lattice site, $h_{LL'}(L \neq L')$ represents the hopping coefficient from the L' th lattice site to the Lth lattice site, and $g_{LL'}$ represents the pairing coefficient of creating or annihilating two phase space phonons on the Lth and L' th lattice sites. These coefficients are given by the phase space force matrix

$$\begin{cases} h_{LL'} \equiv \frac{1}{2}\left(\eta_{XX}^{LL'} + \eta_{PP}^{LL'}\right) + i\frac{1}{2}\left(\eta_{PX}^{LL'} - \eta_{XP}^{LL'}\right) \\ g_{LL'} \equiv \frac{1}{2}\left(\eta_{XX}^{LL'} - \eta_{PP}^{LL'}\right) + i\frac{1}{2}\left(\eta_{PX}^{LL'} + \eta_{XP}^{LL'}\right). \end{cases} \tag{C.8}$$

From property $\eta_{\alpha\beta}^{LL'} = \eta_{\beta\alpha}^{L'L}$, we have $h_{LL'} = h_{L'L}^*$ and $g_{LL'} = g_{L'L}$.

Below, we calculate the explicit form of $h_{LL'}$ and $g_{LL'}$ for the case of Coulomb-type interaction $U(R_{LL'}) = \gamma\pi^{-1}R_{LL'}^{-1}$. For the off-site ($L \neq L'$) coefficients, we have the hopping coefficients

$$\begin{aligned} h_{LL'} &= -\frac{U'(R_{LL'})}{R_{LL'}} - \frac{1}{2}\frac{U''(R_{LL'})R_{LL'} - U'(R_{LL'})}{R_{LL'}} \\ &= -\frac{1}{2}\frac{U''(R_{LL'})R_{LL'} + U'(R_{LL'})}{R_{LL'}} \\ &= -\frac{\gamma}{2\pi R_{LL'}^3} \end{aligned} \tag{C.9}$$

and the pairing coefficients

$$\begin{aligned} g_{LL'} &= -\frac{1}{2}\frac{U''(R_{LL'})R_{LL'} - U'(R_{LL'})}{R_{LL'}^3}[(X_L - X_{L'})^2 - (P_L - P_{L'})^2 \\ &\quad + i2(X_L - X_{L'})(P_L - P_{L'})] \\ &= -\frac{1}{2}\frac{U''(R_{LL'})R_{LL'} - U'(R_{LL'})}{R_{LL'}}e^{i2\varphi_{LL'}} \\ &= -\frac{3\gamma}{2\pi R_{LL'}^3}e^{i2\varphi_{LL'}}, \end{aligned} \tag{C.10}$$

where the phase parameter $\varphi_{LL'}$ is defined via

$$X_L - X_{L'} = R_{LL'}\cos\varphi_{LL'}, \quad P_L - P_{L'} = R_{LL'}\sin\varphi_{LL'}. \tag{C.11}$$

For the on-site ($L = L'$) coefficients, we have the on-site energy

$$\begin{aligned} h_L &= \frac{1}{2}(T_{XX}^{LL} + T_{PP}^{LL}) - \frac{1}{2}\sum_{L'\neq L}(\Phi_{XX}^{LL'} + \Phi_{PP}^{LL'}) \\ &= \frac{1}{2}(T_{XX}^{LL} + T_{PP}^{LL}) + \frac{1}{2}\sum_{L'\neq L}\frac{U''(R_{LL'})R_{LL'} + U'(R_{LL'})}{R_{LL'}} \\ &= \frac{1}{2}(T_{XX}^{LL} + T_{PP}^{LL}) + \sum_{L'\neq L}\frac{\gamma}{2\pi R_{LL'}^3} \end{aligned} \tag{C.12}$$

and the squeezing rates

$$g_L = \frac{1}{2}(T_{XX}^{LL} - T_{PP}^{LL}) + iT_{PX}^{LL} + \sum_{L'\neq L}\frac{3\gamma}{2\pi R_{LL'}^3}e^{i2\varphi_{LL'}}. \tag{C.13}$$

The above expressions are valid for any boundary condition and arbitrary single-particle Hamiltonian. Finally, we obtain the Hamiltonian of phase space lattice waves

$$\begin{aligned} \frac{\mathcal{H}}{\lambda} &= \sum_L \omega_L a_L^\dagger a_L + \frac{1}{2}g_L a_L^{\dagger 2} + \frac{1}{2}g_L^* a_L^2 \\ &\quad - \frac{\gamma}{4\pi}\left(\sum_{L\neq L'}\frac{a_L^\dagger a_{L'} + 3e^{i2\varphi_{LL'}}a_L^\dagger a_{L'}^\dagger}{|\mathbf{Z}_L^0 - \mathbf{Z}_{L'}^0|^3} + h.c.\right). \end{aligned} \tag{C.14}$$

For the honeycomb single-particle Hamiltonian (A.8) and periodic boundary conditions, we have the simplified results of on-site coefficients

$$h_L = \omega_0 \pm 2\Delta + \sum_{L' \neq L} \frac{\gamma}{2\pi R_{LL'}^3}, \quad g_L = 0, \tag{C.15}$$

where the summation in equation (C.13) disappears due to the honeycomb lattice symmetry. The Hamiltonian equation (C.14) describes phonons propagating in our honeycomb phase space crystal. It is reminiscent of the tight-binding model for electrons in graphene but our phase space phonons can hop between any two arbitrarily distant sites reflecting the long-range nature of the atom-atom interaction.

C.2 Pairing-induced hopping

In order to better distinguish the effects of the (long-range) hopping and the pairing interaction, it is instructive to first consider a regime where the latter is suppressed, i.e., the creation of phonon pairs is far off-resonant $\gamma \ll \omega_0$. In this case, we can cancel the pairing-interaction terms in the Hamiltonian (C.7) using the following unitary transformation

$$U \equiv \exp\left(\sum_{\bar{L},\overline{L'}} \xi_{\bar{L}\overline{L'}} a_{\bar{L}}^\dagger a_{\overline{L'}}^\dagger - \xi_{\overline{L'}\bar{L}}^* a_{\bar{L}} a_{\overline{L'}} \right) \equiv e^Y. \tag{C.16}$$

It can be checked that $U^\dagger = e^{-Y}$ and thus $U^\dagger U = 1$. Using the following identities

$$\begin{aligned}
\left[a_L^\dagger a_{L'}, a_{\bar{L}}^\dagger a_{\overline{L'}}^\dagger\right] &= a_L^\dagger a_{\overline{L'}}^\dagger \delta_{L'\bar{L}} + a_L^\dagger a_{\bar{L}}^\dagger \delta_{L'\overline{L'}} \\
[a_L^\dagger a_{L'}, a_{\bar{L}} a_{\overline{L'}}] &= -a_{\overline{L'}} a_{L'} \delta_{L\bar{L}} - a_{\bar{L}} a_{L'} \delta_{L\overline{L'}} \\
[a_L^\dagger a_{L'}^\dagger, a_{\bar{L}} a_{\overline{L'}}] &= -a_L^\dagger a_{\overline{L'}} \delta_{L'\bar{L}} - a_L^\dagger a_{\bar{L}} \delta_{L'\overline{L'}} - a_{\overline{L'}} a_{L'}^\dagger \delta_{L\bar{L}} - a_{\bar{L}} a_{L'}^\dagger \delta_{L\overline{L'}} \\
\left[a_L a_{L'}, a_{\bar{L}}^\dagger a_{\overline{L'}}^\dagger\right] &= a_L a_{\bar{L}}^\dagger \delta_{L'\overline{L'}} + a_L a_{\overline{L'}}^\dagger \delta_{L'\bar{L}} + a_{\overline{L'}}^\dagger a_{L'} \delta_{L\bar{L}} + a_{\bar{L}}^\dagger a_{L'} \delta_{L\overline{L'}},
\end{aligned} \tag{C.17}$$

we obtain

$$\begin{aligned}
[a_L^\dagger a_{L'}, Y] &= \sum_{\bar{L}} \tilde{\xi}_{\bar{L}L'} a_L^\dagger a_{\bar{L}}^\dagger + \tilde{\xi}_{\bar{L}L}^* a_{\bar{L}} a_{L'}, \\
[a_L^\dagger a_{L'}^\dagger, Y] &= \sum_{\bar{L}} \tilde{\xi}_{\bar{L}L'}^* a_L^\dagger a_{\bar{L}} + \tilde{\xi}_{\bar{L}L}^* a_{\bar{L}} a_{L'}^\dagger, \\
[a_L a_{L'}, Y] &= \sum_{\bar{L}} \tilde{\xi}_{\bar{L}L'} a_L a_{\bar{L}}^\dagger + \tilde{\xi}_{\bar{L}L} a_{\bar{L}}^\dagger a_{L'},
\end{aligned} \tag{C.18}$$

where we have defined $\tilde{\xi}_{\bar{L}L'} \equiv \xi_{\bar{L}L'} + \xi_{L'\bar{L}}$ with $\tilde{\xi}_{\bar{L}L'} = \tilde{\xi}_{L'\bar{L}}$. The Hamiltonian transformed by U to the leading order is given by

$$U^\dagger \mathcal{H} U/\lambda = \mathcal{H}/\lambda + [\mathcal{H}/\lambda, Y] + \cdot + \cdots. \tag{C.19}$$

where we have used the Hausdorff expansion

$$e^{-Y}\mathcal{H}e^{Y} = \mathcal{H} + [\mathcal{H}, Y] + \frac{1}{2!}[[\mathcal{H}, Y], Y] + \cdots.$$

The leading-order correction is

$$\begin{aligned}[\mathcal{H}/\lambda, Y] = & \sum_{L;L';\bar{L}} h_{LL'}\xi_{\bar{L}L'}a_{ls}^{\dagger}a_{\bar{L}}^{\dagger} + h_{LL'}\xi_{\bar{L}L}^{*}a_{\bar{L}}a_{L'} \\ & + \frac{1}{2}g_{LL'}\tilde{\xi}_{\bar{L}L'}^{*}a_{L}^{\dagger}a_{\bar{L}} + \frac{1}{2}g_{LL'}\tilde{\xi}_{\bar{L}L}^{*}a_{\bar{L}}a_{L'}^{\dagger} + \frac{1}{2}g_{L'L}^{*}\tilde{\xi}_{\bar{L}L'}a_{L}a_{\bar{L}}^{\dagger} \\ & + \frac{1}{2}g_{L'L}^{*}\tilde{\xi}_{\bar{L}L}a_{\bar{L}}^{\dagger}a_{L'} \\ = & \sum_{L;L'} \tilde{h}_{LL'}a_{L}^{\dagger}a_{L'} + \frac{1}{2}\tilde{g}_{LL'}a_{L}^{\dagger}a_{L'}^{\dagger} + \frac{1}{2}\tilde{g}_{L'L}^{*}a_{L}a_{L'}.\end{aligned} \tag{C.20}$$

Neglecting constants from $a_L a_L^{\dagger} = a_L^{\dagger}a_L + 1$ and using the symmetry property $g_{LL'} = g_{L'L}$, we have the coefficients

$$\begin{aligned}\tilde{h}_{LL'} &= \sum_{\bar{L}} g_{\bar{L}L}^{*}\tilde{\xi}_{\bar{L}L} + g_{L\bar{L}}\tilde{\xi}_{\bar{L}L'}^{*}, \\ \tilde{g}_{LL'} &= \sum_{\bar{L}} h_{L\bar{L}}\tilde{\xi}_{\bar{L}L'}, \\ \tilde{g}_{L'L}^{*} &= \sum_{\bar{L}} h_{\bar{L}L'}\tilde{\xi}_{\bar{L}L}^{*}.\end{aligned} \tag{C.21}$$

To cancel the pairing terms in the Hamiltonian (C.7), it needs $g_{LL'} + \tilde{g}_{LL'} = 0$, i.e.,

$$g_{LL'} = -\sum_{\bar{L}} h_{L\bar{L}}\tilde{\xi}_{\bar{L}L'}. \tag{C.22}$$

We make the assumption that the on-site energy is much larger than the off-site hopping terms, i.e., $\omega_L \equiv h_{LL} \gg h_{L\bar{L}}$ with $L \neq \bar{L}$, which results in

$$\tilde{\xi}_{LL'} \approx -\frac{g_{LL'}}{\omega_L} \quad \Longrightarrow \quad \xi_{LL'} = \xi_{L'L} = -\frac{g_{LL'}}{2\omega_L}.$$

Finally, we have the pairing-induced hopping rate

$$\tilde{h}_{LL'} \approx -\sum_{\bar{L}} \frac{2}{\omega_{\bar{L}}} g_{L\bar{L}}g_{\bar{L}L'}^{*}. \tag{C.23}$$

C.3 Periodic boundary condition

For periodic boundaries both in X and P directions, we Fourier transform the quantum Hamiltonian (C.7) using

$$a_{l,s} = \frac{1}{\sqrt{N}}\sum_{\boldsymbol{k}} a_{\boldsymbol{k},s}\exp(i\boldsymbol{k}\cdot\boldsymbol{Z}_{l,s}^{0});\ a_{l,s}^{\dagger} = \frac{1}{\sqrt{N}}\sum_{\boldsymbol{k}} a_{\boldsymbol{k},s}^{\dagger}\exp(-i\boldsymbol{k}\cdot\boldsymbol{Z}_{l,s}^{0}), \tag{C.24}$$

where $\boldsymbol{k} = (k_X, k_P)$, $Z^0_{l,s}$ is the equilibrium position of atom $L = (l, s)$ and N is the total number of unite cells. The operators from inverse Fourier transformation

$$a_{\boldsymbol{k},s} = \frac{1}{\sqrt{N}} \sum_l a_{l,s} \exp(-i\boldsymbol{k} \cdot \boldsymbol{Z}^0_{l,s}); \qquad a^\dagger_{\boldsymbol{k},s} = \frac{1}{\sqrt{N}} \sum_l a^\dagger_{l,s} \exp(i\boldsymbol{k} \cdot \boldsymbol{Z}^0_{l,s}), \quad \text{(C.25)}$$

satisfy the commutation of Bosons, i.e., $\boldsymbol{k}$

$$\left[a_{\boldsymbol{k},s}, a^\dagger_{\boldsymbol{k}',s'}\right] = \delta_{\boldsymbol{k}\boldsymbol{k}'}\delta_{ss'}; \quad \left[a_{\boldsymbol{k},s}, a_{\boldsymbol{k}',s'}\right] = 0. \quad \text{(C.26)}$$

The hopping terms in the Hamiltonian (C.7) become

$$\begin{aligned} \sum_{l,l'} h_{(l,s),(l',s')} a^\dagger_{l,s} a_{l',s'} &= \frac{1}{N} \sum_{\boldsymbol{k},\boldsymbol{k}'} a^\dagger_{\boldsymbol{k},s} a_{\boldsymbol{k}',s'} \sum_{l,l'} h_{(l,s),(l',s')} e^{-i\boldsymbol{k}\cdot(\boldsymbol{Z}^0_{l,s}-\boldsymbol{Z}^0_{l',s'})-i(\boldsymbol{k}-\boldsymbol{k}')\cdot\boldsymbol{Z}^0_{l',s'}} \\ &\equiv \sum_{\boldsymbol{k}} h_{s,s'}(\boldsymbol{k}) a^\dagger_{\boldsymbol{k},s} a_{\boldsymbol{k},s'} \end{aligned} \quad \text{(C.27)}$$

with the definition of hopping coefficients in reciprocal $\boldsymbol{k} = (k_X, k_P)$ space

$$h_{s,s'}(\boldsymbol{k}) \equiv \sum_l h_{(l,s),(l',s')} \exp[-i\boldsymbol{k} \cdot (\boldsymbol{Z}^0_{l,s} - \boldsymbol{Z}^0_{l',s'})]. \quad \text{(C.28)}$$

We find the property $h^*_{s,s'}(\boldsymbol{k}) = h_{s',s}(\boldsymbol{k})$ using the identity $h_{(l,s),(l',s')} = h^*_{(l',s'),(l,s)}$ and the discrete translational lattice symmetry. Similarly, the pairing terms in the Hamiltonian (C.7) become

$$\begin{aligned} \sum_{l,l'} g_{(l,s),(l',s')} a^\dagger_{l,s} a^\dagger_{l',s'} &= \frac{1}{N} \sum_{\boldsymbol{k},\boldsymbol{k}'} a^\dagger_{\boldsymbol{k},s} a^\dagger_{\boldsymbol{k}',s'} \sum_{l,l'} g_{(l,s),(l',s')} e^{-i\boldsymbol{k}\cdot(\boldsymbol{Z}^0_{l,s}-\boldsymbol{Z}^0_{l',s'})-i(\boldsymbol{k}+\boldsymbol{k}')\cdot\boldsymbol{Z}^0_{l',s'}} \\ &= \sum_{\boldsymbol{k}} g_{s,s'}(\boldsymbol{k}) a^\dagger_{\boldsymbol{k},s} a^\dagger_{-\boldsymbol{k},s'} \end{aligned} \quad \text{(C.29)}$$

with the definition of pairing coefficient in reciprocal $\boldsymbol{k} = (k_X, k_P)$ space

$$g_{s,s'}(\boldsymbol{k}) \equiv \sum_l g_{(l,s),(l',s')} \exp[-i\boldsymbol{k} \cdot (\boldsymbol{Z}^0_{l,s} - \boldsymbol{Z}^0_{l',s'})]. \quad \text{(C.30)}$$

We find the property $g_{s,s'}(-\boldsymbol{k}) = g_{s',s}(\boldsymbol{k})$ using the identity $g_{(l,s),(l',s')} = G_{(l',s'),(l,s)}$ and the discrete lattice translational symmetry. As a result, the Hamiltonian in the reciprocal space is given by

$$\begin{aligned} \mathcal{H}/\lambda &= \sum_{\boldsymbol{k},s,s'} h_{s,s'}(\boldsymbol{k}) a^\dagger_{\boldsymbol{k},s} a_{\boldsymbol{k},s'} + \frac{1}{2} g_{s,s'}(\boldsymbol{k}) a^\dagger_{\boldsymbol{k},s} a^\dagger_{-\boldsymbol{k},s'} + \frac{1}{2} g^*_{s,s'}(\boldsymbol{k}) a_{-\boldsymbol{k},s'} a_{\boldsymbol{k},s} \\ &= \sum_{\boldsymbol{k}} A^T_{\boldsymbol{k}} \mathcal{H}_{\boldsymbol{k}} A_{\boldsymbol{k}}, \end{aligned} \quad \text{(C.31)}$$

where we have denied the vector $A_{\boldsymbol{k}} \equiv (a_{\boldsymbol{k},A}, a_{\boldsymbol{k},B}, a^\dagger_{-\boldsymbol{k},A}, a^\dagger_{-\boldsymbol{k},B})^T$ and the Bogoliubov–de Gennes (BdG) Hamiltonian

$$\mathcal{H}_k \equiv \begin{pmatrix} h_{A,A}(\boldsymbol{k}) & h_{A,B}(\boldsymbol{k}) & \frac{1}{2}g_{A,A}(\boldsymbol{k}) & \frac{1}{2}g_{A,B}(\boldsymbol{k}) \\ h_{B,A}(\boldsymbol{k}) & h_{B,B}(\boldsymbol{k}) & \frac{1}{2}g_{B,A}(\boldsymbol{k}) & \frac{1}{2}g_{B,B}(\boldsymbol{k}) \\ \frac{1}{2}g^*_{A,\,A}(\boldsymbol{k}) & \frac{1}{2}g^*_{A,\,B}(\boldsymbol{k}) & h_{A,A}(-\boldsymbol{k}) & h_{A,B}(-\boldsymbol{k}) \\ \frac{1}{2}g^*_{B,\,A}(\boldsymbol{k}) & \frac{1}{2}g^*_{B,\,B}(\boldsymbol{k}) & h_{B,A}(-\boldsymbol{k}) & h_{B,B}(-\boldsymbol{k}) \end{pmatrix}. \tag{C.32}$$

From the Heisenberg equation

$$\begin{aligned} -i\frac{d}{dt}a_{\boldsymbol{k},s} &= \frac{1}{\lambda}\left[\mathcal{H},\, a_{\boldsymbol{k},s}\right] \\ &= \sum_{s'} -h_{s,s'}(\boldsymbol{k})a_{\boldsymbol{k},s'} - \frac{1}{2}g_{s,s'}(\boldsymbol{k})a^\dagger_{-\boldsymbol{k},\,s'} - \frac{1}{2}g_{s',s}(-\boldsymbol{k})a^\dagger_{-\boldsymbol{k},\,s'}, \end{aligned} \tag{C.33}$$

we have the equation of motion (EOM)

$$i\frac{d}{dt}\boldsymbol{A}_k = D(\boldsymbol{k})\boldsymbol{A}_k. \tag{C.34}$$

Here, the dynamical matrix is defined by

$$D(\boldsymbol{k}) \equiv \begin{pmatrix} h_{A,A}(\boldsymbol{k}) & h_{A,B}(\boldsymbol{k}) & \bar{g}_{A,A}(\boldsymbol{k}) & \bar{g}_{A,B}(\boldsymbol{k}) \\ h_{B,A}(\boldsymbol{k}) & h_{B,B}(\boldsymbol{k}) & \bar{g}_{B,A}(\boldsymbol{k}) & \bar{g}_{B,B}(\boldsymbol{k}) \\ -\bar{g}^*_{A,\,A}(\boldsymbol{k}) & -\bar{g}^*_{B,\,A}(\boldsymbol{k}) & -h^*_{A,\,A}(-\boldsymbol{k}) & -h^*_{A,\,B}(-\boldsymbol{k}) \\ -\bar{g}^*_{A,\,B}(\boldsymbol{k}) & -\bar{g}^*_{B,\,B}(\boldsymbol{k}) & -h^*_{B,\,A}(-\boldsymbol{k}) & -h^*_{B,\,B}(-\boldsymbol{k}) \end{pmatrix} \tag{C.35}$$

with the symmetric pairing coefficients defined by

$$\bar{g}_{s,s'}(\boldsymbol{k}) \equiv \frac{1}{2}\left[g_{s,s'}(\boldsymbol{k}) + g_{s',s}(-\boldsymbol{k})\right].$$

C.4 Strip-shaped boundary condition

For the periodic boundary in X direction but open boundary in P direction (strip-shaped boundary condition), we can only perform the Fourier transformation in the X direction

$$\begin{aligned} a_{l,s} &= \frac{1}{\sqrt{N_X}}\sum_{k_X} a_{k_X,(l_P,s)} \exp(ik_X X^0_{l,\,s}); \\ a^\dagger_{l,\,s} &= \frac{1}{\sqrt{N_X}}\sum_{k_X} a^\dagger_{k_X,\,(l_P,\,s)} \exp(-ik_X X^0_{l,\,s}), \end{aligned} \tag{C.36}$$

where N_X is the number of unit cells in X direction and (l_P, s) labels the position of atoms in P direction. Then, the hopping terms in the Hamiltonian (C.7) become

$$\begin{aligned}\sum_{l,l'} h_{(l,s),(l',s')} a^{\dagger}_{l,\,s} a_{l',s'} &= \frac{1}{N_X} \sum_{l,l'} \sum_{k_X,k'_X} a^{\dagger}_{k_X,\,(l_P,\,s)} a_{k'_X,(l'_P,s')} h_{(l,s),(l',s')} \\ &\qquad e^{-ik_X(X^0_{l,s}-X^0_{l',s'})-i(k_X-k'_X)X^0_{l',s'}} \\ &= \sum_{l_P,l'_P} \sum_{k_X} h_{(l_P,s),(l'_P,s')}(k_X) a^{\dagger}_{k_X,\,(l_P,\,s)} a_{k_X,(l'_P,s')}\end{aligned} \tag{C.37}$$

with the Fourier transformation of hopping coefficients

$$h_{(l_P,s),(l'_P,s')}(k_X) \equiv \sum_{l_X} h_{(l,s),(l',s')} \exp[-ik_X \cdot (X^0_{l,\,s} - X^0_{l',\,s'})], \tag{C.38}$$

which has the property $h^*_{(l'_P,\,s'),\,(l_P,\,s)}(k_X) = h_{(l_P,s),(l'_P,s')}(k_X)$ using the identity $h_{(l,s),(l',s')} = h^*_{(l',\,s'),\,(l,\,s)}$ and discrete translational symmetry in the X direction. Similarly, the pairing terms in the Hamiltonian (C.7) become

$$\begin{aligned}\sum_{l,l'} g_{s,s'}(l-l') a^{\dagger}_{l,\,s} a^{\dagger}_{l',\,s'} &= \frac{1}{N_X} \sum_{l,l'} \sum_{k_X,k'_X} a^{\dagger}_{k_X,\,(l_P,\,s)} a^{\dagger}_{k'_X,\,(l'_P,\,s')} g_{(l,s),(l',s')} \\ &\qquad e^{-ik_X(X^0_{l,s}-X^0_{l',s'})-i(k_X+k'_X)X^0_{l',s'}} \\ &= \sum_{l_P,l'_P} \sum_{k_X} g_{(l_P,s),(l'_P,s')}(k_X) a^{\dagger}_{k_X,\,(l_P,\,s)} a^{\dagger}_{-k_X,\,(l'_P,\,s')}\end{aligned} \tag{C.39}$$

with the Fourier transformation of pairing coefficients

$$g_{(l_P,s),(l'_P,s')}(k_X) \equiv \sum_{l_X} g_{(l,s),(l',s')}(l-l') \exp[-ik_X \cdot (X^0_{l,\,s} - X^0_{l',\,s'})], \tag{C.40}$$

which has the property of $g_{(l_P,s),(l'_P,s')}(-k_X) = g_{(l'_P,s'),(l_P,s)}(k_X)$ using the identity $g_{(l,s),(l',s')} = g_{(l',s'),(l,s)}$ and discrete translational symmetry in the X direction. Therefore, the Hamiltonian in momentum space is

$$\begin{aligned}\mathcal{H}/\lambda = \sum_{l_P,l'_P} \sum_{k_X} & h_{(l_P,s),(l'_P,s')}(k_X) a^{\dagger}_{k_X,\,(l_P,\,s)} a_{k_X,(l'_P,s')} \\ &+ \frac{1}{2} g_{(l_P,s),(l'_P,s')}(k_X) a^{\dagger}_{k_X,\,(l_P,\,s)} a^{\dagger}_{-k_X,\,(l'_P,\,s')} \\ &+ \frac{1}{2} g^*_{(l_P,\,s),\,(l'_P,\,s')}(k_X) a_{k_X,(l_P,s)} a_{-k_X,(l'_P,s')}.\end{aligned} \tag{C.41}$$

From the Heisenberg equation

$$\begin{aligned}-i\frac{d}{dt} a_{k_X,(l_P,s)} = \sum_{l'_P,s'} & - h_{(l_P,s),(l'_P,s')}(k_X) a_{k_X,(l'_P,s')} \\ &- \frac{1}{2}\Big[g_{(l_P,s),(l'_P,s')}(k_X) + g_{(l'_P,s'),(l_P,s)}(-k_X)\Big] a^{\dagger}_{-k_X,\,(l'_P,\,s')},\end{aligned} \tag{C.42}$$

we have the EOM for the operator $A_{l_p}(k_X) \equiv (a_{k_X,(l_P,A)}, a_{k_X,(l_P,B)}, a^{\dagger}_{-k_X,\,(l_P,\,A)}, a^{\dagger}_{-k_X,\,(l_P,\,B)})^T$

$$i\frac{d}{dt}\begin{pmatrix} A_1 \\ \vdots \\ A_{l'_P} \\ \vdots \\ A_{N_P} \end{pmatrix} = \begin{pmatrix} D_{(l_P=1,l'_P=1)} & \cdots & D_{(l_P=1,l'_P)} & \cdots & D_{(l_P=1,l'_P=N_P)} \\ \vdots & \ddots & \vdots & \vdots & \vdots \\ D_{(l_P,l'_P=1)} & \cdots & D_{l'_P,l_P} & \cdots & D_{(l_P,l'_P=N_P)} \\ \vdots & \vdots & \vdots & \ddots & \vdots \\ D_{(l_P=N_P,l'_P=1)} & \cdots & D_{(l_P=N_P,l'_P)} & \cdots & D_{(l_P=N_P,l'_P=N_P)} \end{pmatrix}\begin{pmatrix} A_1 \\ \vdots \\ A_{l'_P} \\ \vdots \\ A_{N_P} \end{pmatrix}, \tag{C.43}$$

where the block matrix element is given by

$$D_{(l_P,l'_P)}(k_X) \equiv \begin{pmatrix} h_{(l_P,A),(l'_P,A)}(k_X) & h_{(l_P,A),(l'_P,B)}(k_X) & \bar{g}_{(l_P,A),(l'_P,A)}(k_X) & \bar{g}_{(l_P,A),(l'_P,B)}(k_X) \\ h_{(l_P,B),(l'_P,A)}(k_X) & h_{(l_P,B),(l'_P,B)}(k_X) & \bar{g}_{(l_P,B),(l'_P,A)}(k_X) & \bar{g}_{(l_P,B),(l'_P,B)}(k_X) \\ -\bar{g}^*_{(l_P,\,A),\,(l'_P,\,A)}(-k_X) & -\bar{g}^*_{(l_P,\,A),\,(l'_P,\,B)}(-k_X) & -h^*_{(l_P,\,A),\,(l'_P,\,A)}(-k_X) & -h^*_{(l_P,\,A),\,(l'_P,\,B)}(-k_X) \\ -\bar{g}^*_{(l_P,\,B),\,(l'_P,\,A)}(-k_X) & -\bar{g}^*_{(l_P,\,B),\,(l'_P,\,B)}(-k_X) & -h^*_{(l_P,\,B),\,(l'_P,\,A)}(-k_X) & -h^*_{(l_P,\,B),\,(l'_P,\,B)}(-k_X) \end{pmatrix} \tag{C.44}$$

with the modified pairing coefficient

$$\bar{g}_{(l_P,s),(l'_P,s')}(k_X) \equiv \frac{1}{2}\left[g_{(l_P,s),(l'_P,s')}(k_X) + g_{(l'_P,s'),(l_P,s)}(-k_X)\right]. \tag{C.45}$$

www.ingramcontent.com/pod-product-compliance
Ingram Content Group UK Ltd.
Pitfield, Milton Keynes, MK11 3LW, UK
UKHW051851100926
8556IPUK00005B/578

9 780750 335614